CISM COURSES AND LECTURES

The series presents lecture notes, monographs, edited works and proceedings in the field of Mechanics, Engineering, Computer Science and Applied Mathematics.
Purpose of the series in to make known in the international scientific and technical community results obtained in some of the activities organized by CISM, the International Centre for Mechanical Sciences.

INTERNATIONAL CENTRE FOR MECHANICAL SCIENCES

COURSES AND LECTURES - No. 343

COMPUTERIZED SYMBOLIC MANIPULATION IN MECHANICS

EDITED BY

E. KREUZER
TECHNICAL UNIVERSITY OF HAMBURG-HARBURG

SPRINGER - VERLAG WIEN - NEW YORK

Le spese di stampa di questo volume sono in parte coperte da
contributi del Consiglio Nazionale delle Ricerche.

This volume contains 64 illustrations

Printed in Italy

In order to make this volume available as economically and as rapidly as possible the authors' typescripts have been reproduced in their original forms. This method unfortunately has its typographical limitations but it is hoped that they in no way distract the reader.

ISBN 3-211-82616-5 Springer Verlag Wien-New York

PREFACE

Computer modeling plays already an important role in the development of new cars or trains, in the design of bridges and light-weight structures and many other mechanical systems. Since the early days computers were seen as machines for calculating numbers in order to draw the right conclusions from the models used to describe reality.

Consequently, computers are now thoroughly established as devices that can greatly ease the human burden of work. We also have access to numerical software libraries and powerful simulation packages which allow the animation of even complex systems. But it is less generally appreciated that also symbols and formulas can be automatically manipulated by symbolic mathematics computing systems or so-called computer algebra systems.

With general-purpose systems we can undertake a wide variety of routine mathematical work which could never be done by paper and pencil. Such mathematical tools enhance the productivity of any scientist and engineer by increasing efficiency and reliability.

Symbolic computation has been an important tool for researchers already for some time and has generated considerable interest in the area of mechanics. The International Centre for Mechanical Sciences in Udine, Italy, has organized a course on "Computerized Symbolic Manipulation in Mechanics" in 1993 in which advanced programs and theories for automatically doing mathematics have been discussed. The lectures have been presented by scientists from Germany and the United States of America; the almost 40 participants came from 12 different countries.

The papers collected in this volume are the outgrowth of the lectures presented during the course. The material is selfcontained and does not require advanced knowledge in symbol manipulation techniques.

The book should be of value for engineers, physicists, and applied mathematicians interested in the computer analysis of complex mechanical systems and phenomena.

The support of Prof. Tasso, editor of this series, is greatly appreciated.

Edwin Kreuzer

CONTENTS

Page

GENERATION OF SYMBOLIC EQUATIONS OF MOTION OF MULTIBODY SYSTEMS

E.J. Kreuzer
Technical University Hamburg-Harburg, Hamburg, Germany

Abstract

In applied mathematics and mechanical engineering we normally think of numerical calculations if computers are involved. Although simulations by means of numerical methods are powerful tools for investigations in mechanics, they do have drawbacks, e.g. finite precision, errors generated when evaluating expressions. General global results or proofs of theoretical results cannot be obtained from simulations.

A broader understanding of mechanical phenomena can be gained by means of analytical methods. But even for seemingly simple mathematical models, analytical calculations by paper and pencil may become very time-consuming, may be the source of many errors, and will sometimes be impossible. In such cases, computerized symbolic manipulation is clearly faster as well as safer and therefore preferable. But often, purely symbolical investigations cannot fulfill all of our needs in mechanics. Therefore, a semi-analytical approach, combining the features of analytical and numerical computations, is a most desirable synthesis. This allows the analytic work to be pushed further before numerical computations start.

The aim of the course is to present important software tools, basic concepts, methods, and applications of computerized symbolic manipulation to mechanics problems. A survey on different approaches and possibilities for symbol manipulation is followed by a review of all major general purpose software packages. Lectures on generation of equations of motion, structural mechanics in general, analysis of nonlinear dynamic systems, and perturbation methods represent some of the most advanced applications of symbol manipulation methods. The main objective of this course is to encourage the use of symbolic manipulation in the analysis of mechanical systems.

1 Overview on Computerized Symbolic Manipulation in Mechanics

In this introductory chapter, we discuss various aspects of computerized symbolic manipulation and some of the capabilities that prove useful when performing computer algebra computations in mechanics.

1.1 Introduction

Digital computers are commercially available since the early fifties of this century. Since then the digital computer has gone through a number of generations of hardware development with concurrent development of languages and systems, Davis (1978). The first true stored-program computers were primarily engaged in numerical calculations, and the early computer literature is full of novel numerical methods, Fitch (1993). Here we look at the part played by the computer which receives increasing interest as a result of the recognition of the central role of algorithms in computer science: the performance of calculus and algebraic operations, especially as required in the physical sciences and engineering.

For numerical calculations the digital computer is now thoroughly established as a device that can greatly ease the human burden of work. In these computations, the quantities which are manipulated are numbers, either integer or floating point. This allows a wide range of problems to be solved, and has become known as scientific computing.

Especially in the analysis of mechanical systems numerical methods became important and in fact, the availability of relatively inexpensive computing power has stimulated much of the research in mechanics. Moreover, results obtained from numerical simulations inspired researchers to develop theories which help to explain phenomena that have not been understood for a long time. The reason for using computers in numerical experiments is simple: computers can relatively easily calculate solutions of mechanical systems whereas for only a few nonlinear systems we can formulate closed-form solutions. This is in contrast to linear systems, where closed-form solutions can be obtained in terms of eigenvalues and eigenvectors. Numerical methods are often generally applicable but even if the result of a simulation would be exact, it is just one solution of one system from one initial condition using one set of parameter values, Parker and Chua (1989). Hence, although simulations by means of numerical methods are a powerful tool for investigations in mechanics, they do have their limitations. Numerical procedures use finite precision arithmetic and are based on the theory of approximation, therefore, errors generated when evaluating expressions cannot be avoided. Moreover, general global results or proofs of theoretical results cannot be obtained from numerical simulations. They do, however, suggest ideas that lead to proofs. An example is Appel and Haken's computer proof of the four-color theorem.

A broader understanding of system's structure and physical phenomena can be gained by means of analytical methods. The lack of closed-form solutions for nonlinear equations could lead one to believe, that at least approximate solutions are feasible. But in order to determine periodic solutions, their stability, and bifurcations from them, analytical calculations by paper and pencil may become very time-consuming, as well as error-prone and sometimes even impossible. Approximations of nonlinear models by higher order series expansions for multiple-dimensional systems may also be a formidable task. But with improvements in technology, and with non-numerical applications attracting growing attention, all this is changing. In many cases, symbol manipulation by means of computer algebra systems is of considerable help. A symbolic algorithm pays, however, for the exactness with – in general much – higher execution time and storage requirements compared to its numerical counterpart, Loos (1982).

Since their invention, computer algebra systems always operated at their limit with respect to memory capacity and time. This situation stimulated the development of algorithms e.g. for dealing with problems such as arithmetic of rational functions consisting of several thousands of terms. But already at the beginning of the use of computers for symbolic manipulation it became apparent that the straightforward methods were often very inefficient.

Neither purely numerical nor purely symbolical calculations will fulfill our needs in mechanics. For the study of mechanical systems, a semi-analytical approach combining the features of analytical and numerical approaches is, therefore, a desirable approach. Thereby, analytical methods use results of previous numerical calculations such as coefficients of polynomial approximations; contrarily analytically generated equations might need to be solved numerically.

1.2 Overview on Symbol Manipulation

The modern digital computer is a general-purpose machine that can carry out any algorithm or precisely specified procedure. Internally, numerical mathematics may be used as part of the application, but numbers are not in general the immediately desired results. Thus, an airline reservation system provides a variety of information including availability of seats for a given flight, which obviously involves some simple arithmetic, Davis (1978). Non-numerical applications of the computer have an enormous impact on the average person. The number of non-numerical users of computers is today very large and continues to increase.

It is less generally appreciated, however, that there are computer programs equally well adapted to the manipulation of algebraic expressions: the computer can work not only with numbers but also with more abstract symbols that represent numerical quantities.

With symbol manipulation or computer algebra we nowadays describe the branch of science which uses computers to do algebraic calculations rather than to perform purely arithmetic calculations which we all know computers can do. These calculations may relate to many areas of mathematics such as algebra, calculus, or geometry.

The first documented use of computerized symbol manipulation for formal differentiation dates back to the early fifties, Nolan (1953), Kharimanian (1954). But ever since, the use of computers in natural sciences and engineering for numerical calculations was much broader compared with non-numerical applications. This was caused mainly by the lack of appropriate software. Since the mid-sixties and especially in the eighties, better applicable and easier accessible general-purpose software for symbol manipulation has been developed.

The early systems were often produced by scientists with particular problems to solve and some have grown into the major systems of today. While there have been significant mathematical and computational advances in the subject, the origins were in applications, and applications still form the most important part of computer algebra, Fitch (1993).

The first systems were driven by large problems such as classical mechanics, general relativity and quantum mechanics. Computer algebra was expensive with respect to machine resources and it only seemed worthwhile constructing systems for such large problems. Big problems in celestial mechanics required sometimes months of calculation by hand. By means of a computer algebra system the computer confirmed the results in minutes. It is important to note, however, that methods used in computer algebra are often well-suited for machine calculation but less convenient by hand. This is a common feature of computer algebra, the methods used are not always the familiar ones. The details of these algorithms are beyond the scope of this chapter, they usually use advanced mathematical skills.

A feature of all the larger systems is the calculation of a Gröbner basis from a set of polynomials. This apparently obscure algorithm is possibly the biggest algorithmic advance in computer algebra in 20 years. It can be used to handle side-relations in calculations, solve sets of nonlinear equations, attack major geometrical problems, and many other things.

As in the case of celestial mechanics, there is now a growing number of scientists who would never consider performing extensive calculations by hand with the risk of losing a sign or a factor, when a machine can ensure accuracy: computer algebra is their servant.

For a long time computer algebra was a specialist subject for people with large computers. But increases in processor speed and cheaper computer memory mean that computer algebra systems and users no longer require the largest computers. Most systems are now available for most workstations and there are several systems for personal computers.

1.3 Current Systems

At present, there are a number of general-purpose computerized symbol manipulation systems in use. Without trying to give an exhaustive list we will mention only the most important software with a large variety of functions. These are the systems REDUCE, MACSYMA, MAPLE, MATHEMATICA, and AXIOM. The developers of each system chose a certain set of abilities they wanted their programs to have. The development of the systems REDUCE and MACSYMA, both written in LISP, started already in the sixties. These two are some of the first commercially available interactive symbol manipulation programs. These well-established and reliable programs have a rich collection of symbol manipulation tools. Libraries contain routines ranging from simplest trigonometric identities to complex tensor calculations. Both systems are nowadays available on a large variety of machines.

The vast improvements in computer technology and the decrease in hardware costs encouraged the development of new symbol manipulation software like MAPLE and MATHEMATICA, both written in portable C code. The graphic capabilities of MAPLE are highly developed and it was among the first packages to provide a smooth transition between algebraic and numerical calculation. Another important contribution to the symbol manipulation software is the system AXIOM. This software provides many of the most advanced algorithms but is presently restricted to a few specific types of computers. Although e.g. MACSYMA is considerably slower than MAPLE and MATHEMATICA, its strength is its extensive library of well-tested routines and functions. There are also other systems such as MATLAB and MATHCAD which provide some symbolic capabilities without being true algebraic systems.

Computer algebra systems offer a large variety of options to solve a wide range of mathematical problems. Some items are listed below:

- Algebra

  Complex arithmetic, exact and floating point arithmetic, trigonometric and hyperbolic functions and their inverses, substitution and part extraction, rank, determinant, inversion and triangularization, etc.

- Calculus

  Differentiation, partial derivatives of unknown functions, limits, Taylor and Laurent series, indefinite and definite integrals, Laplace and Fourier transforms, etc.

- Vector and Tensor Analysis

  Vector calculus, tensor calculus, etc.

Furthermore, some systems provide additional features like

- Graphics
  Two-dimensional plots, three-dimensional plots, contour plots, hardcopies, etc.
- Utilities
  Programmability, output in FORTRAN and C code, interface to numerical programs, LaTeX-output, etc.

Table 1.1 summarizes important features of some computer algebra systems and provides a partial assessment and comparison of the capabilities of these systems. Any comparison depends, however, very much on what kind of calculation we wish to perform, on the equipment, and personal preferences. A completely objective review is impossible, for each person has a different level of competency with computers and has different amounts of experience with programming.

Table 1.1: Summary of features of various computer algebra systems

criterion	**program**			
	MACSYMA	MATHEMATICA	MAPLE	AXIOM
programming language	LISP	C	C	LISP
quality of algorithms	++	+−	++	+++
user interface	+	+++	++	+++
FORTRAN output	•	•	•	•
UNIX facility	∘	•	•	•
machine dependency	∘	∘	∘	•
CPU time	−−	++	+	+
costs	−	+	+++	−−−

+/− : positive/negative performance
• : feature available
∘ : feature not available

Scientists who are concerned with solving problems in their own field would, naturally, prefer to employ an existing system to satisfy their needs. It has been demonstrated with general-purpose software like MACSYMA and REDUCE, that they can be of great help in various fields of mechanics, Davis (1978), Noor and Andersen (1979), Zeis and Cefola (1980). The system MACSYMA has built up a corpus of applications over the years, with particular strengths, for example, in preparing finite element codes.

Experience has revealed, however, a number of drawbacks to such general purpose systems. If only a very specific problem has to be solved like the evaluation of Lagrange's equations for a complicated dynamic system resulting in the explicit equations of motion, than a special-purpose program may handle the task better. A general-purpose system, however, is written to handle a large variety of mathematical contingencies, many of which are rarely required. The generality therefore results in code which occupies computer storage, thereby reducing its availability to the problem solver and hence leading to inefficiency for the specific problem solution. It happens quite often, that available computer space will be exceeded, thus making it impossible to continue. In the course of analytical calculations usually an unpredictable amount of space is required for expressions which are generated. Simplification techniques for cancellation of terms may reduce expressions, but their size may still exceed available capacity. The following example may illustrate, that seemingly simple problems can result in calculations no longer treatable by existing software, Wolfram (1988).

Finding eigenvalues of an $n \times n$-matrix in general involves solving an n-th degree polynomial equation. Since for $n \geq 5$ explicit algebraic solutions cannot in general be found, it is impossible to give explicit algebraic results for the eigenvalues and eigenvectors of all but the simplest or sparsest matrices.

Presently, the computational efficiency has become less critical than some years ago and one could, therefore, seriously consider the use of general-purpose systems in place of special-purpose programs. But careful comparisons showed, that still today special-purpose programs for a restricted number of mathematical operations and functions are faster, require less memory space, and may be more efficient. One problem in mechanics for which the non-numeric application of computers was early recognized is the generation of equations of motion of multibody systems. We will provide an extensive discussion of this application of symbol manipulation within this course.

1.4 Contents of the Course

The purpose of this CISM-Course is to provide a coherent presentation of symbolic and algebraic computation tools and techniques that support the investigation of mechanical systems. The course brings together scientists engaged in research of computerized symbolic manipulation from different academic institutions to present

- Computer Algebra Software for Scientific Application,
- Generation of Symbolic Equations of Motion of Multibody Systems,

- Computerized Symbolic Computation in Structural Mechanics,
- Perturbation Methods and Computer Algebra in Mechanics,
- Analysis of Nonlinear Dynamic Engineering Systems.

Computerized symbolic manipulation has such power and a great variety of applications in mechanics that we hope people will begin using such systems and start exploring those parts of problems that could not otherwise be studied analytically.

2 Modeling of Engineering Systems

Analysis and simulation of engineering systems is based upon appropriate models of real systems. In mechanical engineering the development of a mathematical model is subdivided into two steps: (1) forming a mechanical model of the system under consideration, (2) generating the mathematical model by using physical laws. The first step is discussed in this chapter, the second step is described in various lectures within this course. The methods developed are applied to an industrial robot. Examples are distributed troughout the text to demonstrate important aspects.

2.1 Introduction

Growing requirements of speed and precision make it necessary to operate a technical system closer to the dynamic limits rather than to overdesign a system and work too far away from the mechanical possibilities. These increasing demands have motivated progress in the analysis and simulation of complex mechanical systems. The development of mathematical models of mechanical systems, however, requires both knowledge of the best methods of mechanics as well as practical experience. Mathematical modeling is like an art. The basic tools can be formally learned, but the rest is mostly acquired through experience. Recipes or "how-to" guidelines are not easily taught or described in books. In general, the better the behavior of a system is understood the simpler the mathematical model that may be employed.

The impact of the computer on the engineering sciences has broadened the domain of the mathematical modeler, who can now reliably simulate quite complex physical phenomena by computer. Widely used in all sciences and in engineering, and a research area in its own right, computer modeling plays a major role in the development of many critical technologies such as satellite and aircraft dynamics as well as in vehicle, robot and machine dynamics. Developing appropriate simulation tools for a given technology invariably involves a high degree of scientific knowledge and sophisticated mathematical tools to describe and evaluate the models. Validation of these models may require not only statistical tests but also experimental comparisons. Moreover, the model itself has to be tuned to physical data to confirm or improve its aptness for representing a physical process or phenomenon.

Our aim is to present within this course various symbol manipulation tools for a systematic and efficient analysis of mechanical systems. A strategy for the analysis of nonlinear dynamic systems is shown in Fig. 2.1. This strategy is equally applicable for mechanical systems in general.

The determination of the response of complex mechanical systems to dynamic loads is essential for design and evaluation of system's performance under varying conditions before such systems are constructed. But what kind of mechanical model should be

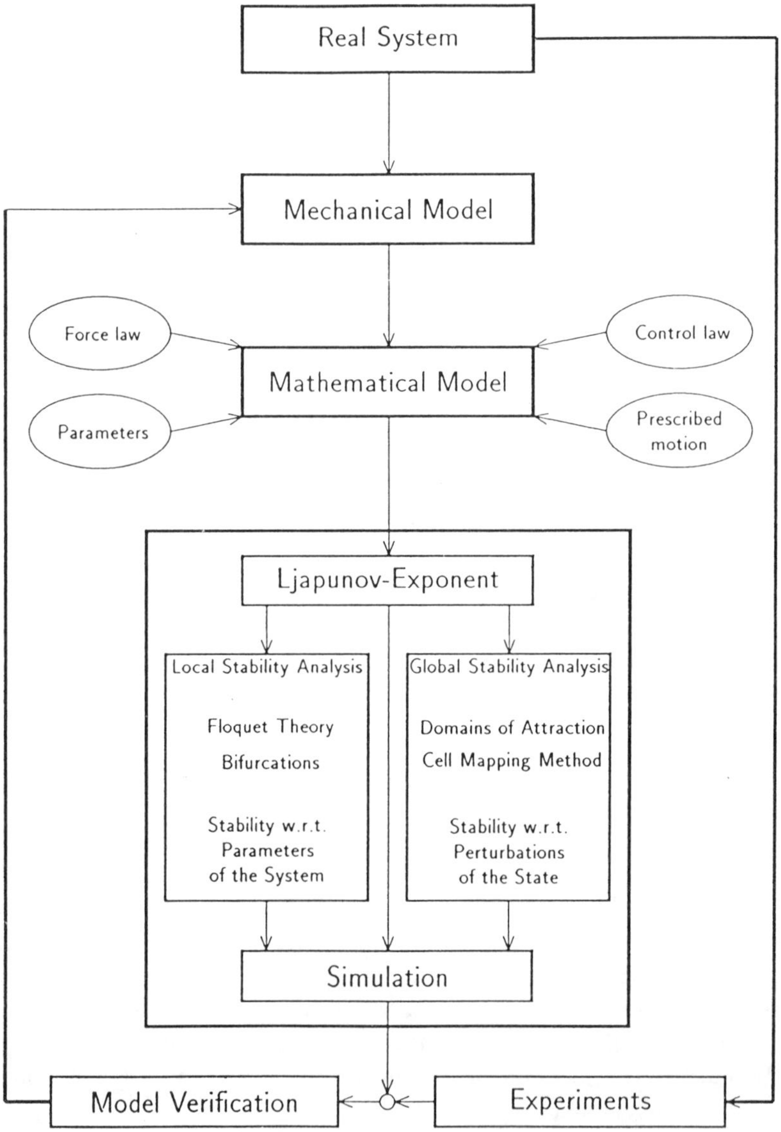

Figure 2.1: Flowchart for the analysis of nonlinear dynamic systems

used or is most adequate for the actual problem? At the present time the practicing engineer may use four different kinds of mechanical models:

- Multibody Systems,

- Finite Element Systems,
- Continuous Systems,
- Hybrid Multibody Systems.

The complexity of the models increases in the mentioned order. Depending on the complexity of the system under consideration, it may be conceptually advisable to use a series of mechanical models with increasing refinements. Starting with the simplest or crudest model, refinements are added in successive steps. However, the general guideline in mechanical modeling always should be: devise a model as simple as possible. It is the combination of theoretical knowledge and practical experience, which leads to well-suited models, and which is often so difficult to reach.

2.2 Multibody Systems

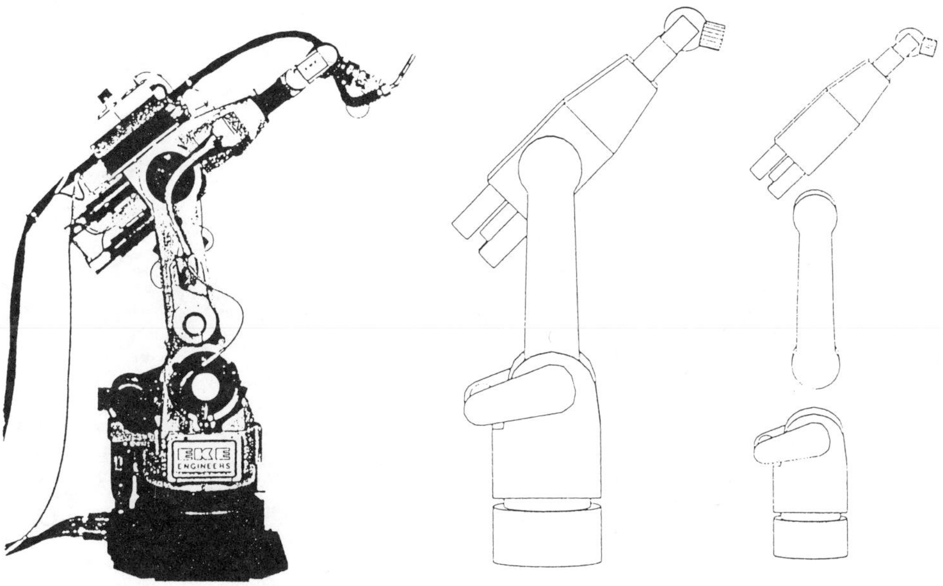

Figure 2.2: Multibody system model of an industrial robot

Engineering systems, for example industrial robots, usually consist of several structural elements. The elements are often stiff and can therefore be modeled as rigid bodies, coupled by inelastic joints. The essential effects on the dynamics caused by the large rigid body motions are well reproduced by *M*ulti*B*ody *S*ystems (MBS). The number of resulting equations of motion remains comparatively small, and this makes the multibody system approach very appropriate for control system design.

Multibody systems have become very popular in various fields of engineering because they often provide more physical insight than can be obtained from more complicated models. Furthermore, the label multibody system is usually reserved for systems which undergo large motions such that the dynamics cannot be adequately described by linear differential equations. But there are many real systems, in many disciplines of mechanical engineering, which can at least approximately be described as linear multibody systems, Fig. 2.2.

The simplest modeling of elastic structural elements is to equally distribute mass and elasticity of an extended body into lumped masses, interconnected by massless springs and dampers. The additional abilities of movement increase the number of degrees of freedom of these so-called lumped mass systems. A more accurate mechanical model of an engineering system with elastic structural elements is obtained by joint beam elements. In contrast to lumped mass systems a beam is built up by elements, consisting of at least three flexibly connected rigid bodies which are additionally coupled by springs, Fig. 2.3. A detailed analysis and discussion of elastic beam systems was presented by Rauh (1987).

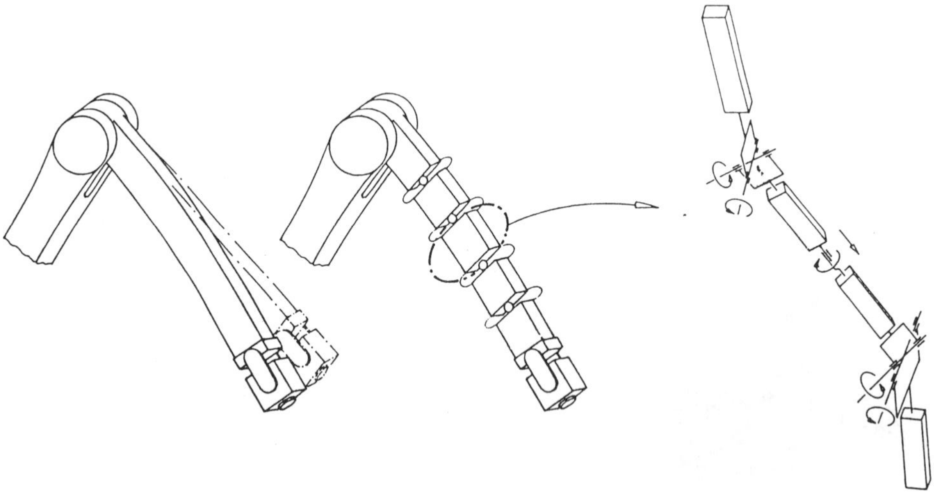

Figure 2.3: Joint-beam element for modeling beam structures

Multibody models of elastic structural elements are characterized by a rather small number of degrees of freedom, but, on the other hand, this approach does lead to problems and limits:

- For a good approximation of the elastic behavior it is necessary to define springs which overlap several bodies. The determination of the corresponding spring coefficients is complicated and requires a good insight into the expected system motion.

- The modeling of three-dimensional deformations and especially the coupling of different elementary cases of deformation is difficult.
- Simple lumped mass systems in general reproduce only the first fundamental mode shape correctly, but fundamental mode shapes of higher order may often not be neglected. Useful results are only obtained for bending strains in a definite preferred direction.

2.3 Finite Element Systems

Finite element methods belong to the main tools of structural engineering analysis. The finite element technique, which has been traditionally used as a purely numerical one, can provide analytical results, when performed with the help of a symbolic manipulation code.

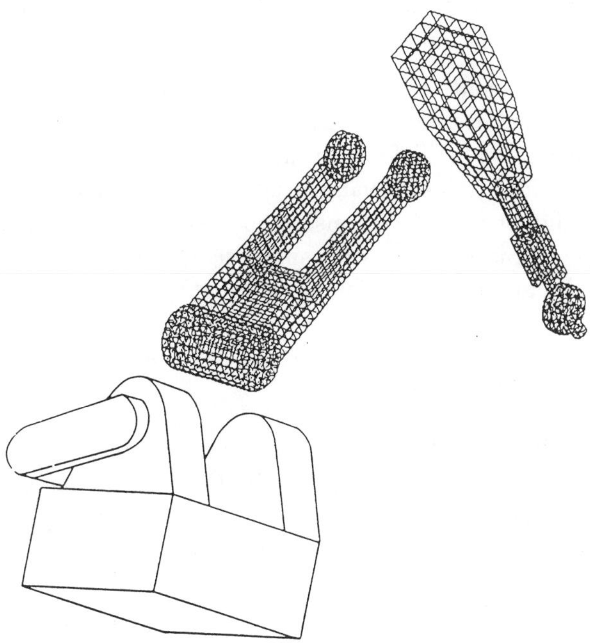

Figure 2.4: Finite element model of an industrial robot

*F*inite *E*lement *S*ystems (FES) are convenient for the discretization of engineering systems with elastic structural elements, because there are several standard procedures for the modeling of complex structures, Fig. 2.4. The modeling of engineering systems is subdivided into three steps:

1. Division of the structure into homogeneously distributed elements (when possible), so that all geometric boundary and compatibility conditions may be fulfilled by the node displacement.

2. Investigation of the particular elements, whose behavior is described only by the displacements at the nodes.

3. Combination of the elements to describe the entire system.

Finite element systems are appropriate for the analysis of static loads and small dynamic motions referred to an inertial system. The superposition of large motions of coincident points (rigid body motions) and small elastic motions causes difficulties and is not considered in corresponding software systems. Sunanda and Dubowsky (1982) included results of a finite element software into the analysis of a multibody system software. This means, in general, that only quasi-static calculations are carried out for many positions of the structural elements. The oscillations during the actual large motions are not taken into account.

Models based on finite element systems lead to many degrees of freedom, with consequential costs of numerical evaluations. There are, however, highly elaborated calculation programs taking advantage of the specific structure of the systems of equations, characterized by sparse matrices with band structure. These mechanical models are not convenient for the design of control systems without an order reduction. Finite element systems, however, may be useful for layout and design, because stress calculations give important hints for design and dimensioning of the structural elements.

2.4 Continuous Systems

Engineering systems with beam-type components are most easily modeled by so-called *CO*ntinuous *S*ystems (COS), Fig. 2.5. The distribution of mass and elasticity is exactly described by these systems. The mathematical formulation of the model leads to a set of partial differential equations, which may be solved exactly only for bodies of particularly simple shape. In general, such systems may only be solved when a separation is possible; that is, an approach that allows the separation of the distributed parameters like mass and elasticity into products of position and time functions.

Due to these problems the method of continuous systems also leads to approximate solutions because of truncation errors. Therefore a "mechanical discretization" by means of joint beam elements and finite elements often is preferable to a variable separation approach.

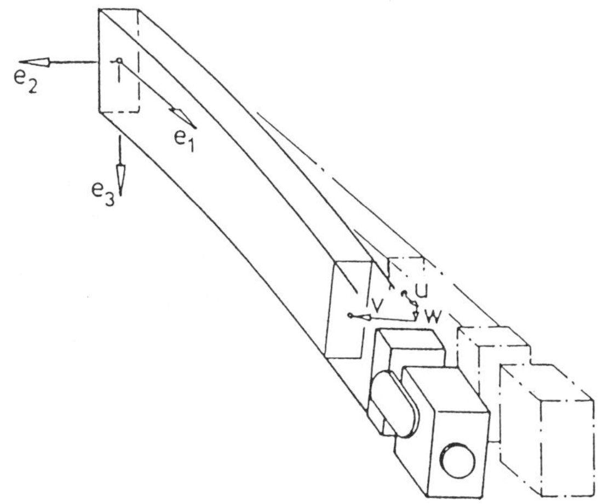

Figure 2.5: Continuous system model of a beam element

2.5 Hybrid Multibody Systems

Mechanical systems with flexible elements are sometimes not satisfactory described by rigid-body-models if the elastic properties have an essential influence on the dynamics. Finite element methods yield good results concerning the deformations and tensions. If we deal with active systems, their high number of degrees of freedom, however, causes difficulties with regard to a reasonable control system design.

Therefore, for engineering systems with elastic structural elements an extension of the method of multibody systems by elastic subbodies is advisable. Such *H*ybrid *M*ulti*B*ody *S*ystems (HMBS) assure a good description of the dynamical behavior combined with a low system order. This allows to embrace torsional and bending strains in all directions, Fig. 2.6. This action combines the advantages of multibody systems and appropriate consideration of small elastic motions.

A hybrid multibody system is a mechanical model consisting of a finite number of rigid bodies and deformable bodies interconnected by massless passive or active elements.

The position and time independent deformation coordinates in every deformation direction are also approximated by a product approach. The number of the degrees of freedom increases with the number of terms considered in the product approach for the position functions. The discretization of the distributed parameters leads to a mathematical description involving ordinary differential equations.

Based on experience, the elastic properties of the structural elements are sufficiently described by at most 3 – 4 terms of the series expansion in each deformation direction. However, this considerably increases the order of the mathematical model causing problems during simulation and, first of all, for the layout of control systems.

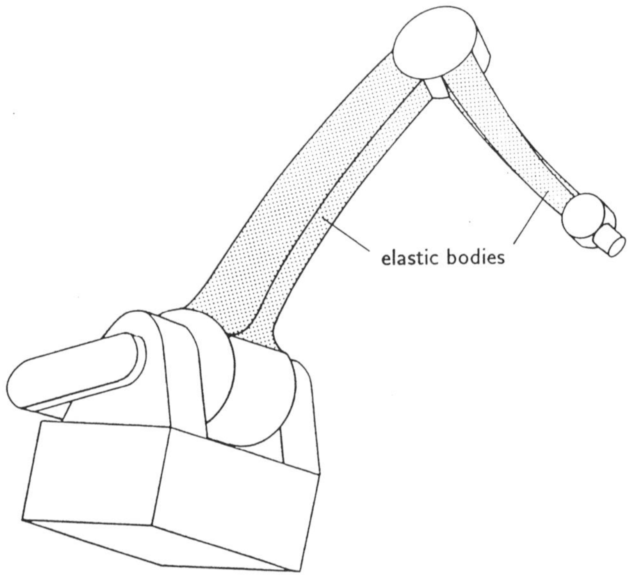

Figure 2.6: Hybrid multibody model of an industrial robot

For the analysis of structural oscillations it suffices to restrict oneself to the effects of the large motions of coincident points.

Hybrid multibody systems lead to the most complex models of engineering systems; however, they possess several advantages:

- most exact description of the distributed parameters and proper interpretation of the results,
- appropriate consideration of spatial deformations and coupling of different deformations (bending and torsion),
- simple identification of parameters like modulus of elasticity and moments of inertia,
- simple consideration of higher order modes of motion by the inclusion of additional terms in the series expansions.

Truckenbrodt (1980) carried out detailed investigations of the influences of different formulations of the solution and the truncation of terms in the corresponding infinite series. Further considerations of hybrid multibody systems can be found in Bremer (1988).

2.6 Comparison of the Mechanical Models

The methods presented in the preceding sections are summarized and classified in Table 2.1. Like all such tables this classification cannot be accurate and extensive. The transition boundaries between the variants of modeling are partly moving. It is essential to note above all, that hybrid multibody systems may be built up of a combination of multibody systems and continuous systems as well as finite element systems. Even a combination of all three methods of modeling seems feasible.

Table 2.1: Classification of the modeling variants

Modeling	HMBS		
	MBS	COS	FES
body	solid	elastic	elastic
geometry	complex	simple	complex
description of deformation	reduced	available	available
forces/moments	discrete	steady distributed	discrete
suitability for design of controling system	good	after reduction	conditional, after reduction

3 Multibody System Dynamics

For dynamic systems, such as vehicles, machinery, and robots, discrete models consisting of rigid bodies, springs, dampers, actuators, and rigid bearings or joints have long proven themselves useful. Such multibody models are called mechanical systems even if they also contain devices which are electric, magnetic, hydraulic or pneumatic in nature.

In this chapter we provide a systematic mathematical description of arbitrary multibody systems. Newton's and Euler's equations are applied to each body. D'Alembert's or Jourdain's principle is used to eliminate reaction forces and torques and to simultaneously reduce system's order to the minimum number of equations necessary.

3.1 Introduction

Dynamical equations governing the behavior of technical devices which can be modeled as systems of interconnected rigid bodies play an important role in the design and operation of such devices. Therefore, the first and surely one of the most important steps in the analysis of engineering dynamic systems is the generation of system's equation of motion.

Two classes of problems arise in which dynamic equations play an important role. In design, these equations are employed to carry out simulations of the motion of dynamic systems such as vehicles, mechanisms, robots, and spacecraft, Fig. 3.1. Here, they furnish a powerful tool for the study of control strategies, for the optimization of parameters, and the testing of a system's performance under varying conditions. In connection with system operation, equations of motion are used for the evaluation of nominal actuator forces and torques in order to achieve a desired motion. Such calculations normally need to be performed on-line; hence, computational efficiency is a major concern.

The methods of analytical dynamics, however, do not often seem to be adequate in order to formulate the equations of motion for complex systems. Even for simple two- or three-body systems, a considerable effort is needed to arrive at Lagrangian equations. On the other hand, for modern design and control systems it is highly recommendable to provide dynamic equations as accurate as possible. Consequently, the equations are often highly nonlinear and are complicated due to the appearance of trigonometric functions.

In this chapter we will rely exclusively on multibody systems. However, having transformed the real system into a multibody system the next step is the generation of the mathematical model. For a multibody system model the mathematical model is a set of ordinary differential equations known as equations of motion. These equations relate

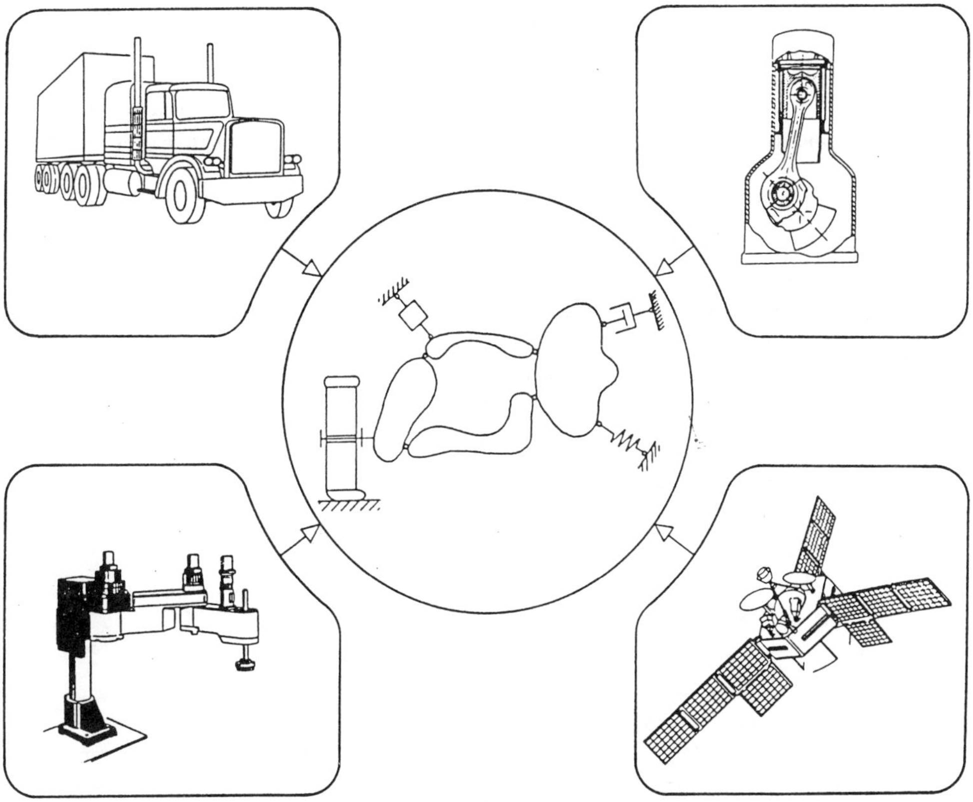

Figure 3.1: Examples for multibody system models

the accelerations to the time, the positions, the velocities, and the parameters of the system.

This requires, especially for complex mechanical systems, a very systematic approach. We might think of Lagrange's method or Newton-Euler's method. However, deriving mathematical models of complicated mechanical systems with paper and pencil and using the best suited method is still very time-consuming and produces unlimited possibilities of errors. Hence, the difficulties in deriving equations of motion efficiently and correctly can prevent us from dealing with complex mechanical systems. These problems led to a desire for a proper multibody simulation code. Therefore, to transfer this tedious task to computers and to generate the equations of motion automatically, a number of formalisms have been developed. Since the early sixties many researches have considered the possibility of computer-aided methods to generate these equations. The basis for these computer-oriented methods is a systematic algorithmic formulation of the dynamical equations.

The increased availability and performance of computer algebra software has made available a powerful tool for constructing programs for symbolic equation generation.

Lagrange's method tends to lead to computational algorithms involving a large number of unnecessary arithmetic operations. Furthermore, the determination of constraint forces is complicated. Consequently, most formalisms use a Newton-Euler approach, Fig. 3.2.

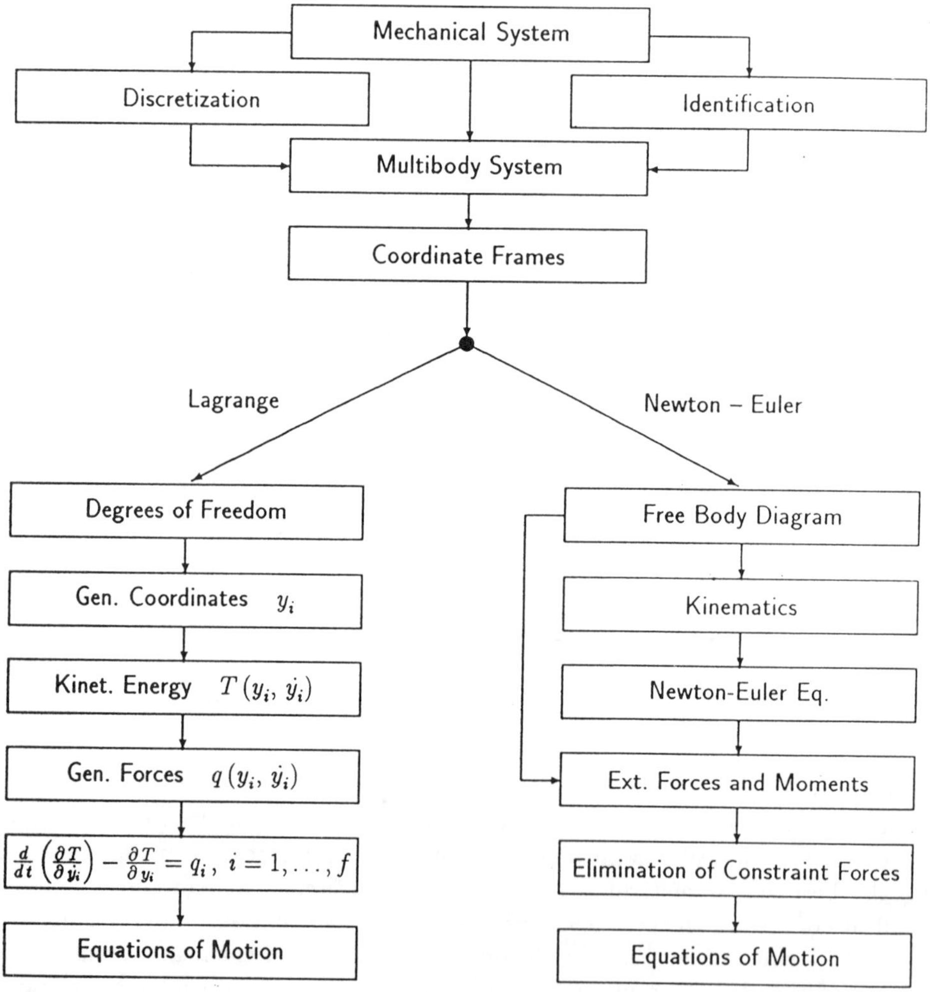

Figure 3.2: Two methods for the derivation of equations of motion

In the following sections we give a systematic description of the kinematics and dynamics of multibody systems.

3.2 Topology and Elements of Multibody Systems

From a topological point of view multibody systems might occur with three different structures, Fig. 3.3:

- open kinematic chain structure,
- tree structure,
- closed loop structure.

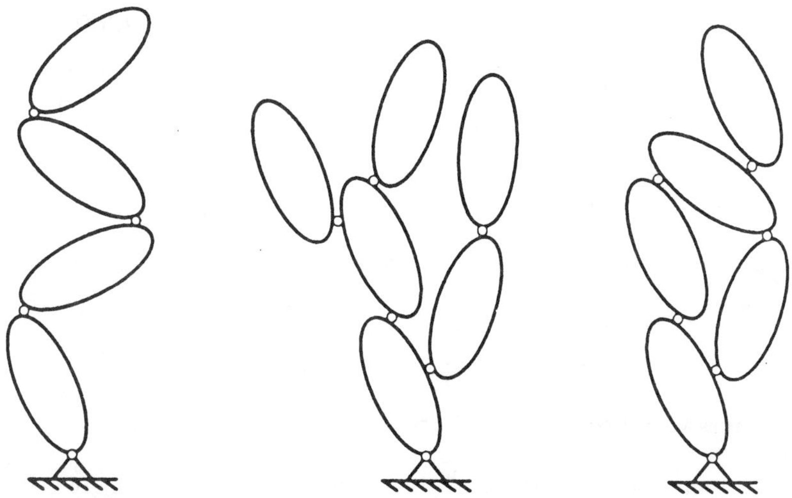

(a) (b) (c)

Figure 3.3: Toplogical structures of multibody systems: (a) chain, (b) tree, (c) closed loop

If a mechanical system consists of connected rigid bodies such that adjacent bodies share at least one common point and such that no closed loops are formed, the system has either open chain or tree structure. Such systems are models of chains, cables, manipulators, robots, antennas, beams, etc.

Characteristic for open chain and tree structures is that the the system is divided into two parts if one cuts a joint between adjacent bodies. For some multibody system formalisms difficulties arise if closed loops or closed kinematic chains are included in the system. These problems appear, because the closing conditions are often not easy to formulate.

Multibody systems with closed loops often occur in mechanism dynamics. Such systems normally result in a set of coupled ordinary differential and algebraic equations. Differential algebraic equations have to be integrated by implicit methods. Numerical experiments by Leister (1992) have shown the integration algorithms for ordinary differential equations being more efficient than solvers for differential-algebraic equations.

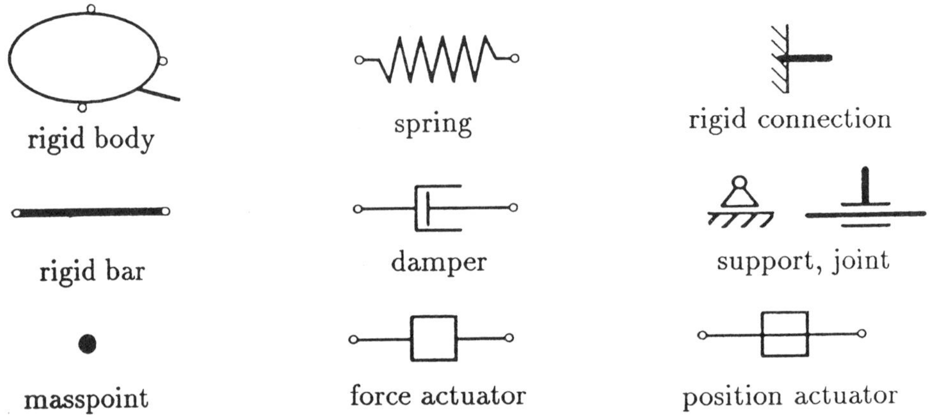

Figure 3.4: Elements of multibody systems

For modeling real systems the method of multibody systems offers the following elements, Fig. 3.4:

- rigid bodies with translational and rotational inertia,
- masspoints with translational inertia only,
- constraining elements without elasticity and friction (such as supports, bearings, Hooke's joints, and rolling wheels),
- coupling elements without inertia acting at nodal points (such as rigid bars, springs, dampers, actuators, and tires).

3.3 Kinematics

Consider a system of p rigid bodies, Fig. 3.5. The bodies are interconnected and may be connected with the environment by bearings or joints. Acting in and/or on the system are passive and active coupling elements such as springs, dampers, and actuators.

A body-fixed frame is associated with each body with the origin at the center of mass. Usually, we may describe the system relative to an inertial reference frame. Then, the position of body i is described by a 3×1–vector $\boldsymbol{r}_i$ to the center of mass C_i and a 3×3–rotation matrix $\boldsymbol{S}_i$ relating the inertial frame and the body–fixed frame. The mass of body i is m_i and its inertia tensor with respect to the center of mass C_i is denoted by $\boldsymbol{I}_i$.

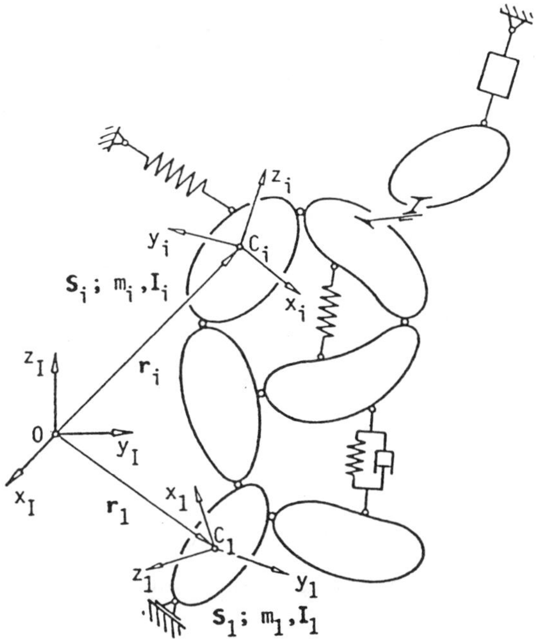

Figure 3.5: Multibody system

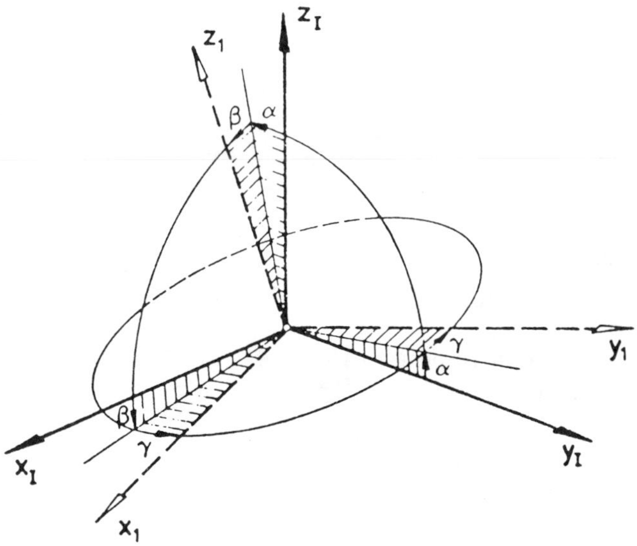

Figure 3.6: Cardan angles α, β, and γ

There are many different ways in order to describe the orientation of a rigid body with respect to an inertial frame or relative to neighboring bodies. One approach is given by means of elementary rotation matrices using cardan angles α, β and γ, Fig. 3.6:

$$\boldsymbol{S}_i^\alpha = \begin{bmatrix} 1 & 0 & 0 \\ 0 & \cos\alpha & -\sin\alpha \\ 0 & \sin\alpha & \cos\alpha \end{bmatrix} , \tag{3.1}$$

$$\boldsymbol{S}_i^\beta = \begin{bmatrix} cos\beta & 0 & \sin\beta \\ 0 & 1 & 0 \\ -\sin\beta & 0 & \cos\beta \end{bmatrix} , \tag{3.2}$$

$$\boldsymbol{S}_i^\gamma = \begin{bmatrix} \cos\gamma & -\sin\gamma & 0 \\ \sin\gamma & \cos\gamma & 0 \\ 0 & 0 & 1 \end{bmatrix} . \tag{3.3}$$

Figure 3.7 shows the relative orientation of adjacent bodies of a robot.

If free rotation with respect to a reference frame is possible the orientation is determined by

$$\boldsymbol{S}_i = \boldsymbol{S}_i^\alpha \boldsymbol{S}_i^\beta \boldsymbol{S}_i^\gamma , \tag{3.4}$$

and results in

$$\boldsymbol{S}_i = \begin{bmatrix} \cos\beta\cos\gamma & -\cos\beta\sin\gamma & \sin\beta \\ \begin{matrix}\cos\alpha\sin\gamma \\ +\sin\alpha\sin\beta\cos\gamma\end{matrix} & \begin{matrix}\cos\alpha\cos\gamma \\ -\sin\alpha\sin\beta\sin\gamma\end{matrix} & -\sin\alpha\cos\beta \\ \begin{matrix}\sin\alpha\sin\gamma \\ -\cos\alpha\sin\beta\cos\gamma\end{matrix} & \begin{matrix}\sin\alpha\cos\gamma \\ +\cos\alpha\sin\beta\sin\gamma\end{matrix} & \cos\alpha\cos\beta \end{bmatrix} . \tag{3.5}$$

The 3×3-inertia tensor with respect to the center of mass C_i may be given in an arbitrary body fixed frame,

$$_i\boldsymbol{I}_i = \begin{bmatrix} I_{xx} & -I_{xy} & -I_{zx} \\ -I_{xy} & I_{yy} & -I_{yz} \\ -I_{zx} & -I_{yz} & I_{zz} \end{bmatrix} , \quad i = 1, \ldots, p , \tag{3.6}$$

or in the principal axis frame

$$_i\boldsymbol{I}_i = \text{diag}\{I_{xx}\ I_{yy}\ I_{zz}\} , \quad i = 1, \ldots, p . \tag{3.7}$$

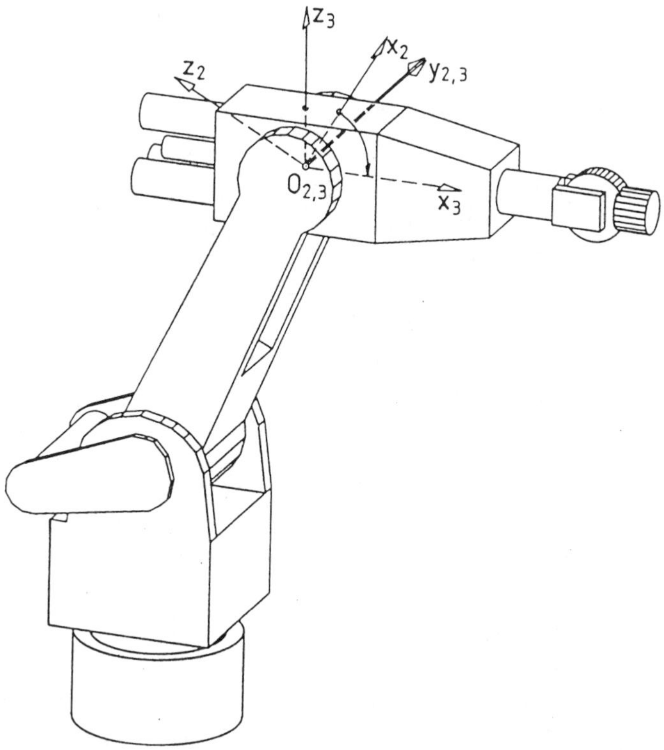

Figure 3.7: Relative orientation of two adjacent bodies

There are a number of applications where a moving reference frame might be more adequate, e. g., in vehicle dynamics. Such moving reference frames are often given in a natural way: a truck running on a curved highway is naturally described with respect to a moving highway-related frame, Fig. 3.8.

The position of each body is again described by a vector

$$\boldsymbol{r}_i = \boldsymbol{r}_R + \boldsymbol{r}_{Ri} \tag{3.8}$$

and a rotation matrix

$$\boldsymbol{S}_i = \boldsymbol{S}_R \boldsymbol{S}_{Ri} \,. \tag{3.9}$$

The reference quantities are denoted by an index R and the relative quantities by the indices Ri. It should be mentioned that the choice of the reference frame does not affect the equations of motion. But the generation process may be considerably simplified.

The constraints imposed by joints, bearings, wheels, etc. may be holonomic or non-holonomic.

Figure 3.8: Definition of reference frames

3.3.1 Holonomic Systems

Holonomic systems are characterized by q holonomic rheonomic constraints. Holonomic constraints are for example given by revolute, prismatic, or spheric joints, Fig. 3.9. Since there are q constraints, the number of degrees of freedom is reduced to $f = 6p - q$. Consequently, the position of the system can be uniquely described by f generalized coordinates, often called Lagrangian coordinates, summarized in an $f \times 1$–vector

$$\boldsymbol{y} = [y_1 \; y_2 \ldots y_f]^T \; . \tag{3.10}$$

Usually, we can find the number f simply by inspection rather than by determining the number of constraints q. The position of the bodies is then described depending on the generalized coordinates:

$$\boldsymbol{r}_i = \boldsymbol{r}_i(\boldsymbol{y}, t) \; , \; \boldsymbol{S}_i = \boldsymbol{S}_i(\boldsymbol{y}, t) \; , \quad i = 1, \ldots, p \; . \tag{3.11}$$

The velocity of the bodies is found by differentation of their position. After a further differentation the absolute acceleration is derived:

$$\dot{\boldsymbol{v}}_i = \boldsymbol{J}_{Ti}(\boldsymbol{y}, t)\ddot{\boldsymbol{y}} + \dot{\bar{\boldsymbol{v}}}_i(\boldsymbol{y}, \dot{\boldsymbol{y}}, t) \; , \quad i = 1, \ldots, p \; , \tag{3.12}$$

$$\dot{\boldsymbol{\omega}}_i = \boldsymbol{J}_{Ri}(\boldsymbol{y}, t)\ddot{\boldsymbol{y}} + \dot{\bar{\boldsymbol{\omega}}}_i(\boldsymbol{y}, \dot{\boldsymbol{y}}, t) \; , \quad i = 1, \ldots, p \; , \tag{3.13}$$

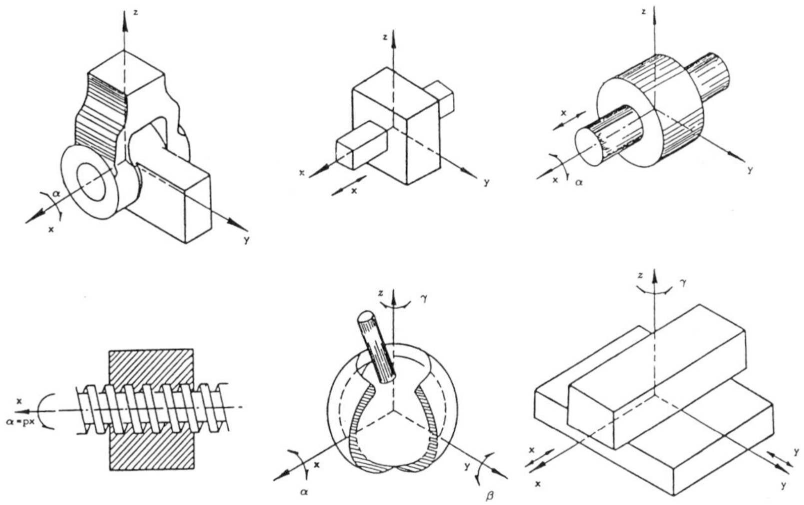

Figure 3.9: Examples of joints resulting in holonomic constraints

The $3 \times f$-Jacobian matrices $\boldsymbol{J}_{Ti}$ and $\boldsymbol{J}_{Ri}$ are specified by the holonomic constraints. The 3×1-vectors $\bar{\dot{\boldsymbol{v}}}_i$ and $\bar{\dot{\boldsymbol{\omega}}}_i$ summarize the remaining terms:

$$\bar{\dot{\boldsymbol{v}}}_i = \frac{\partial \boldsymbol{v}_i}{\partial \boldsymbol{y}} \dot{\boldsymbol{y}} + \frac{\partial \boldsymbol{v}_i}{\partial t} \; , \; \bar{\dot{\boldsymbol{\omega}}}_i = \frac{\partial \boldsymbol{\omega}_i}{\partial \boldsymbol{y}} \dot{\boldsymbol{y}} + \frac{\partial \boldsymbol{\omega}_i}{\partial t} \; . \tag{3.14}$$

The partial derivatives $\frac{\partial \boldsymbol{v}_i}{\partial t}$ and $\frac{\partial \boldsymbol{\omega}_i}{\partial t}$ vanish in case of scleronomic constraints.

3.4 Nonholonomic Systems

These systems are characterized by r nonholonomic rheonomic constraints. An example of a nonholonomic constraint is represented by a rigid wheel rolling on a rough plane. With these additional motion constraints the motional degrees of freedom are reduced to

$$g = f - r \; . \tag{3.15}$$

Thus, the velocity of the system is uniquely described by g generalized velocities summarized in a velocity vector

$$\boldsymbol{z} = [z_1 \; z_2 \ldots z_g]^T \; . \tag{3.16}$$

Of course, generalized velocities may sometimes also be useful in case of holonomic constraints to improve efficiency in the generation process.

Nonholonomic constraints are introduced explicitly by

$$\dot{\boldsymbol{y}} = \boldsymbol{g}'(\boldsymbol{y}, \boldsymbol{z}, t) \ . \tag{3.17}$$

The bodies's velocity is now described in terms of the position vector $\boldsymbol{y}$ and the velocity vector $\boldsymbol{z}$:

$$\boldsymbol{v}_i = \boldsymbol{v}_i(\boldsymbol{y}, \boldsymbol{z}, t) \ , \ \boldsymbol{\omega}_i = \boldsymbol{\omega}_i(\boldsymbol{y}, \boldsymbol{z}, t) \ , \quad i = 1, \ldots, p \ . \tag{3.18}$$

Again, the absolute acceleration is found by differentiation:

$$\dot{\boldsymbol{v}}_i = \boldsymbol{L}_{Ti}(\boldsymbol{y}, \boldsymbol{z}, t)\dot{\boldsymbol{z}} + \dot{\bar{\bar{\boldsymbol{v}}}}_i(\boldsymbol{y}, \boldsymbol{z}, t) \ , \quad i = 1, \ldots, p \ , \tag{3.19}$$

$$\dot{\boldsymbol{\omega}}_i = \boldsymbol{L}_{Ri}(\boldsymbol{y}, \boldsymbol{z}, t)\dot{\boldsymbol{z}} + \dot{\bar{\bar{\boldsymbol{\omega}}}}_i(\boldsymbol{y}, \boldsymbol{z}, t) \ , \quad i = 1, \ldots, p \ , \tag{3.20}$$

where the $3 \times g$-Jacobian kinematical matrices $\boldsymbol{L}_{Ti}$ and $\boldsymbol{L}_{Ri}$ are introduced, and the vectors

$$\dot{\bar{\bar{\boldsymbol{v}}}} = \frac{\partial \boldsymbol{v}_i}{\partial \boldsymbol{y}}\dot{\boldsymbol{y}} + \frac{\partial \boldsymbol{v}_i}{\partial t} \ , \ \dot{\bar{\bar{\boldsymbol{\omega}}}} = \frac{\partial \boldsymbol{\omega}_i}{\partial \boldsymbol{y}}\dot{\boldsymbol{y}} + \frac{\partial \boldsymbol{\omega}_i}{\partial t} \ . \tag{3.21}$$

summarize the remaining terms.

3.5 Dynamics

The application of the linear and angular momentum principle based on the free-body diagram, Fig. 3.10, to each body results in Newton's and Euler's equations

$$\boldsymbol{m}_i\dot{\boldsymbol{v}}_i = \boldsymbol{f}_i^a + \boldsymbol{f}_i^c \ , \quad i = 1, \ldots, p \ , \tag{3.22}$$

$$\boldsymbol{I}_i\dot{\boldsymbol{\omega}}_i + \boldsymbol{\omega}_i \times \boldsymbol{I}_i\boldsymbol{\omega}_i = \boldsymbol{l}_i^a + \boldsymbol{l}_i^c \ , \quad i = 1, \ldots, p \ . \tag{3.23}$$

The external forces and torques are composed of applied forces $\boldsymbol{f}_i^a$ and torques $\boldsymbol{l}_i^a$ and constraint or reaction forces $\boldsymbol{f}_i^c$ and torques $\boldsymbol{l}_i^c$, respectively.

The constraint forces and torques can be described in a very systematic way. If we introduce generalized constraint forces

$$\boldsymbol{g} = [g_1 \ g_2 \cdots g_{q+r}]^T \tag{3.24}$$

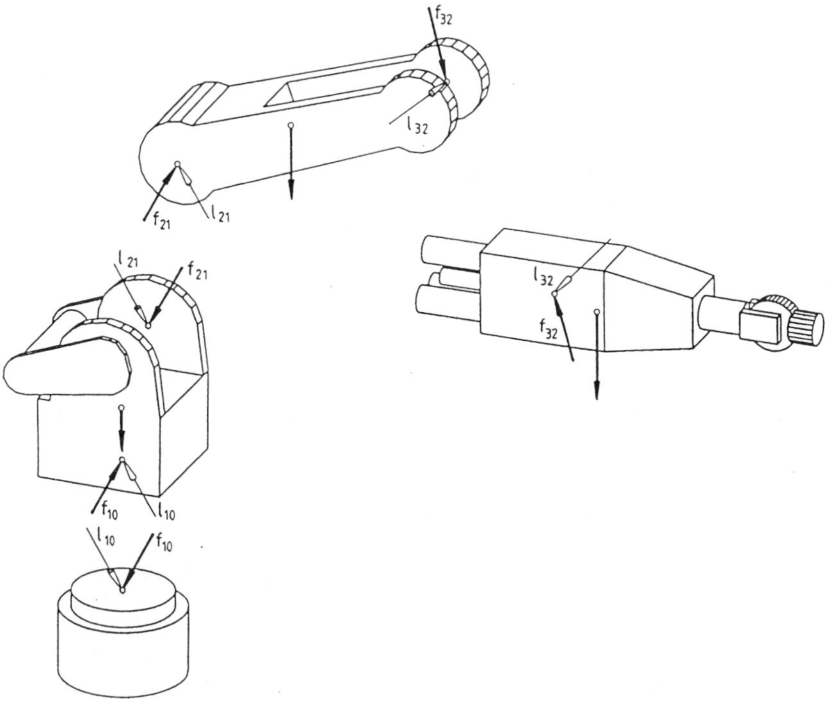

Figure 3.10: Free body diagram of an industrial robot

we can write

$$\boldsymbol{f}_i^c = \boldsymbol{F}_i \boldsymbol{g} \ , \ \boldsymbol{l}_i^c = \boldsymbol{L}_i \boldsymbol{g} \ , \quad i = 1, \dots, p \ , \tag{3.25}$$

where the distribution matrices $\boldsymbol{F}_i$ and $\boldsymbol{L}_i$ are introduced. These matrices follow from geometrical analysis or from implicit constraint equations.

The generalized constraint forces are important design parameters, e. g., for bearings and joints.

To summarize the Newton-Euler equations of all p bodies we introduce global matrices and vectors.

The $6p \times 6p$–block diagonal matrix $\bar{\bar{\boldsymbol{M}}}$ is composed as follows:

$$\bar{\bar{\boldsymbol{M}}} = \text{diag}\{m_1 \boldsymbol{E} \dots m_p \boldsymbol{E} \, \boldsymbol{I}_1 \dots \boldsymbol{I}_p\} \ , \tag{3.26}$$

where $\boldsymbol{E}$ denotes the 3×3–identity matrix. Furthermore, for holonomic and nonholonomic systems we use $6p \times f$–global Jacobian matrices

$$\bar{\boldsymbol{J}} = [\boldsymbol{J}_{T1}^T \dots \boldsymbol{J}_{Tp}^T\, \boldsymbol{J}_{R1}^T \dots \boldsymbol{J}_{Rp}^T]^T , \tag{3.27}$$

$$\bar{\boldsymbol{L}} = [\boldsymbol{L}_{T1}^T \dots \boldsymbol{L}_{Tp}^T\, \boldsymbol{L}_{R1}^T \dots \boldsymbol{L}_{Rp}^T]^T . \tag{3.28}$$

The further quantities on the left hand side of (3.22) and (3.23) are due to gyroscopic effects as well as due to the relations (3.14) and (3.21) and are summarized in $6p \times 1$–vectors $\bar{\boldsymbol{k}}$ and $\bar{\bar{\boldsymbol{k}}}$, respectively. The applied forces and torques are described by the vectors $\bar{\boldsymbol{q}}^a$ and $\bar{\bar{\boldsymbol{q}}}^a$, e. g.,

$$\bar{\boldsymbol{q}}^a = [\boldsymbol{f}_1^{aT} \dots \boldsymbol{f}_p^{aT}\, \boldsymbol{l}_1^{aT} \dots \boldsymbol{l}_p^{aT}]^T , \tag{3.29}$$

whereas for the constraint forces and torques the global distribution matrices $\bar{\boldsymbol{Q}}$ and $\bar{\bar{\boldsymbol{Q}}}$, e. g.,

$$\bar{\boldsymbol{Q}} = [\boldsymbol{F}_1^T \dots \boldsymbol{F}_p^T\, \boldsymbol{L}_1^T \dots \boldsymbol{L}_p^T]^T \tag{3.30}$$

are introduced. With (3.26) to (3.30) the global equation of motion of a multibody system of p bodies is given by

$$\bar{\bar{\boldsymbol{M}}}\bar{\boldsymbol{J}}\ddot{\boldsymbol{y}} + \bar{\boldsymbol{k}}(\boldsymbol{y}, \dot{\boldsymbol{y}}, t) = \bar{\boldsymbol{q}}^a(\boldsymbol{y}, \dot{\boldsymbol{y}}, t) + \bar{\boldsymbol{Q}}(\boldsymbol{y}, t)\boldsymbol{g} \tag{3.31}$$

for holonomic systems and by

$$\bar{\bar{\boldsymbol{M}}}\bar{\boldsymbol{L}}\dot{\boldsymbol{z}} + \bar{\bar{\boldsymbol{k}}}(\boldsymbol{y}, \boldsymbol{z}, t) = \bar{\bar{\boldsymbol{q}}}^a(\boldsymbol{y}, \boldsymbol{z}, t) + \bar{\bar{\boldsymbol{Q}}}(\boldsymbol{y}, \boldsymbol{z}, t)\bar{\boldsymbol{g}} \tag{3.32}$$

for nonholonomic systems or if generalized velocities are introduced in holonomic systems.

The Newton-Euler equations (3.31) or (3.32) are combined algebraic-differential equations. Hence the question arises: Can these equations be separated for solution into purely algebraic and differential equations? We can give a positive answer by employing the principles of dynamics. In a first step, the system's motion can be found by integration of the separated differential equations. Then the equations (3.25) can be solved. For ideal applied forces both steps can be executed successively, the resulting equations of motion are independent of the constraint forces.

3.5.1 Holonomic Systems

For holonomic systems Lagrange's form of D'Alembert's principle states that the virtual work of constraint forces vanishes. This results in

$$\delta W^c = \delta \boldsymbol{y}^T \bar{\boldsymbol{J}}^T \bar{\boldsymbol{Q}}\boldsymbol{g} = 0 ; \tag{3.33}$$

hence, the matrices $\bar{\boldsymbol{J}}^T$ and $\bar{\boldsymbol{Q}}$ are orthogonal. Consequently, pre-multiplication of (3.31) by $\bar{\boldsymbol{J}}^T$ yields a complete elimination of the constraint forces. The result is a second order nonlinear differential equation of motion

$$\boldsymbol{M}(\boldsymbol{y},t)\ddot{\boldsymbol{y}} + \boldsymbol{k}(\boldsymbol{y},\dot{\boldsymbol{y}},t) = \boldsymbol{q}(\boldsymbol{y},\dot{\boldsymbol{y}},t) \ , \tag{3.34}$$

where $\boldsymbol{M}$ is a symmetric and positive definite $f \times f$-inertia matrix, and $\boldsymbol{q}$ and $\boldsymbol{k}$ are $f \times 1$-vectors of generalized applied and gyroscopic forces. Equation (3.34) describes the motion of the system in terms of the minimal number of differential equations. The constraint forces are completely eliminated. Consequently, the vector $\boldsymbol{q}^c$ of constraint forces does not need to be regarded if only the equations of motion are of interest.

The formulation of equations of motion of multibody systems with loops, such as the wheel suspension of Fig. 3.11a, is in general done in two steps. Firstly, the loops are cut and a system with tree-topology is obtained, Fig. 3.11b. The kinematics of this system can be described by independent coordinates. The equations of motion are derived as outlined above and result in equation (3.34). Secondly, the loops are closed by an appropriate number of constraints. If e.g. m implicit algebraic constraints are necessary, the mathematical formulation results in

$$\boldsymbol{c}(\boldsymbol{y},t) = \boldsymbol{0} \ . \tag{3.35}$$

The closed-loop system will then have $f = n - m$ degrees of freedom. The dynamics of the system with loops is described by a set of differential-algebraic equations

$$\boldsymbol{M}^O(\boldsymbol{y},t)\ddot{\boldsymbol{y}} + \boldsymbol{k}^O(\boldsymbol{y},\dot{\boldsymbol{y}},t) - \boldsymbol{q}^O(\boldsymbol{y},\dot{\boldsymbol{y}},t) - \boldsymbol{C}^T\boldsymbol{\lambda} = \boldsymbol{0} \ , \quad \boldsymbol{c}(\boldsymbol{y},t) = \boldsymbol{0} \ , \tag{3.36}$$

where the index ^{O} characterizes the equations of the open system, $\boldsymbol{C} = \frac{\partial \boldsymbol{c}}{\partial \boldsymbol{y}}$ is the Jacobian matrix of the constraints and the Lagrange multipliers $\boldsymbol{\lambda} \in \mathrm{R}^m$ can be regarded as generalized reaction forces in the joints which were cut, Leister and Bestle (1992). In order to arrive at pure differential equations the Lagrange multipliers have to be eliminated and, moreover, the number of the equations of motion can be reduced to the minimal one. This can be performed symbolically and by modifying an integration routine a systematic approach is obtained which automatically chooses a set of independent variables in order to avoid singularities in the loops, Schirm (1993).

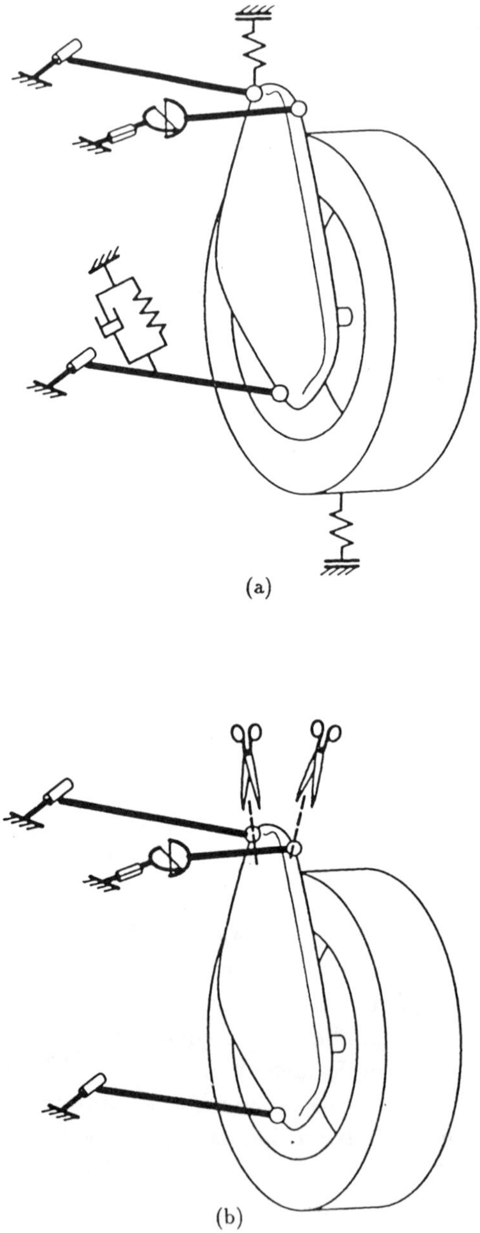

Figure 3.11: Model of a wheel suspension

For linear systems, equation (3.34) can be decomposed as

$$\boldsymbol{M}(t)\ddot{\boldsymbol{y}} + \boldsymbol{P}(t)\dot{\boldsymbol{y}} + \boldsymbol{Q}(t)\boldsymbol{y} = \boldsymbol{h}(t) \ , \tag{3.37}$$

and if the matrices $\boldsymbol{P}$ and $\boldsymbol{Q}$ are time-invariant, even into

$$\boldsymbol{M}(t)\ddot{\boldsymbol{y}} + (\boldsymbol{D} + \boldsymbol{G})\dot{\boldsymbol{y}} + (\boldsymbol{K} + \boldsymbol{N})\boldsymbol{y} = \boldsymbol{h}(t) \ . \tag{3.38}$$

Here $\boldsymbol{P}$ denotes the matrix of velocity dependent forces, $\boldsymbol{D}$ the damping matrix, $\boldsymbol{G}$ the matrix of gyroscopic forces, $\boldsymbol{Q}$ the matrix of position dependent forces, $\boldsymbol{K}$ the stiffness matrix and $\boldsymbol{N}$ the matrix of nonconservative forces. The matrices $\boldsymbol{D}$ and $\boldsymbol{K}$ represent the symmetric, the matrices $\boldsymbol{G}$ and $\boldsymbol{N}$ the skew-symmetric parts of the matrices $\boldsymbol{P}$ and $\boldsymbol{Q}$. The right hand side of equations (3.37) and (3.38) is formed by the time-dependent $f \times 1$-excitation vector $\boldsymbol{h}(t)$.

3.5.2 Nonholonomic Systems

For nonholonomic systems or if generalized velocities are introduced in holonomic systems Jourdain's principle states that the virtual power of all constraint forces and torques vanishes:

$$\delta P^c = \delta \boldsymbol{z}^T \bar{\boldsymbol{L}}^T \bar{\bar{\boldsymbol{Q}}} \boldsymbol{g} = 0 \ . \tag{3.39}$$

Thus, the matrices $\bar{\boldsymbol{L}}^T$ and $\bar{\bar{\boldsymbol{Q}}}$ are orthogonal, and the constraint forces and torques in (3.32) are eliminated by pre-multiplying with $\bar{\boldsymbol{L}}^T$. The result is a first order vector differential equation

$$\boldsymbol{M}^*(\boldsymbol{y}, \boldsymbol{z}, t)\dot{\boldsymbol{z}} + \boldsymbol{k}^*(\boldsymbol{y}, \boldsymbol{z}, t) = \boldsymbol{q}^*(\boldsymbol{y}, \boldsymbol{z}, t) \ . \tag{3.40}$$

Here $\boldsymbol{M}^*$ is again a symmetric and positive definite $g \times g$-inertia matrix and $\boldsymbol{k}^*$ and $\boldsymbol{q}^*$ describe generalized gyroscopic and applied forces. For a complete description of the dynamics the kinematic equations (3.17) must be included.

Frequently, we can obtain useful information about the behavior of a system from linearized equations of motion. The equations of motion (3.34) or (3.40) are linearized by omitting all terms of second and higher order of the generalized coordinates and their derivatives. The vectors $\boldsymbol{k}^*$ and $\boldsymbol{q}^*$ may then be further decomposed.

In case of linear systems the equations of motion (3.40) are decomposed again according to (3.34):

$$\boldsymbol{M}^*(t)\dot{\boldsymbol{z}} + \boldsymbol{P}^*(t)\boldsymbol{z} + \boldsymbol{Q}^*(t)\boldsymbol{y} = \boldsymbol{h}^*(t) \ . \tag{3.41}$$

Further splitting as in (3.38) is not possible in case of nonholonomic systems, though.

3.6 Constraint Forces

Forces and torques that make no contribution to the motion are called constraint forces and constraint torques, respectively. But in case of non-ideal constraints with Coulomb friction or contact problems, the applied forces are coupled with the constraint forces. Hence, the constraint forces occur also in the equations of motion. Especially for design purposes and also for stress estimations, the calculation of constraint forces is necessary. In the following, only holonomic systems are considered.

Rewriting D'Alembert's principle as an orthogonality condition

$$\bar{\boldsymbol{Q}}^T \bar{\bar{\boldsymbol{M}}}^{-1} \bar{\bar{\boldsymbol{M}}} \bar{\boldsymbol{J}} = \boldsymbol{0} \; , \tag{3.42}$$

the elimination of the acceleration is performed through pre-multiplication of (3.31) by $\bar{\boldsymbol{Q}}^T \bar{\bar{\boldsymbol{M}}}^{-1}$. The result is the so-called reaction equation

$$\boldsymbol{N}(\boldsymbol{y},t)\boldsymbol{g} + \hat{\boldsymbol{q}}(\boldsymbol{y},\dot{\boldsymbol{y}},t) = \hat{\boldsymbol{k}}(\boldsymbol{y},\dot{\boldsymbol{y}},t) \; , \tag{3.43}$$

where the $q \times 1$-vectors $\hat{\boldsymbol{q}}$ and $\hat{\boldsymbol{k}}$ characterize the influence of applied and gyroscopic forces, and with the positive definite $q \times q$-reaction matrix $\boldsymbol{N} = \bar{\boldsymbol{Q}}^T \bar{\bar{\boldsymbol{M}}}^{-1} \bar{\boldsymbol{Q}}$, Schramm (1985). Besides using a minimal set of equations this approach has also the advantage that the generalized constraint forces $\boldsymbol{g}$ can be found numerically by standard procedures.

In the case of contact or friction problems, where $\boldsymbol{q}$ also depends on $\boldsymbol{g}$, we have to solve the equations of motion (3.34) and the reaction equations (3.43) simultaneously.

3.7 Inverse Dynamics

The inverse problem of dynamics, namely to find the appropriate drive forces and torques of the actuator in order to perform a desired motion, is often solved using recursive methods. However, recursive equations destroy the structure of the dynamic model which might be useful for controller design.

In the literature control laws are often employed which neglect the complex dynamics, thus leading to less than optimal response. The formalism presented here does not use any approximations and can be considered to be exact under the modeling specifications.

From the equations of motion (3.34) the driving forces are easily determined. Decomposing the vector $\boldsymbol{q}$ of generalized applied forces results in the vector of driving forces

$$\boldsymbol{q}^d = \boldsymbol{M}(\boldsymbol{y})\ddot{\boldsymbol{y}} + \boldsymbol{k}(\boldsymbol{y},\dot{\boldsymbol{y}}) - \boldsymbol{q}^w(\boldsymbol{y},\dot{\boldsymbol{y}}) \tag{3.44}$$

where $\boldsymbol{q}^w$ describes the generalized applied forces due to weight, springs, and actuators and $\boldsymbol{y}$, $\dot{\boldsymbol{y}}$, and $\ddot{\boldsymbol{y}}$ are functions of time.

The driving forces and torques may also be obtained by the constraint force approach outlined in the previous section. But then they can be determined only numerically.

4 Generation of Equations of Motion of Multibody Systems

The dynamics of a multibody system cannot be adequately described by linear differential equations since large motions are characteristic for such systems. The derivation of the equations of motion of such a system is tedious, time-consuming, and error-prone. The process of generating equations of motion involves many vector and matrix manipulations and the generated equations may consist of hundreds of terms. Therefore, already in the sixties, initiated through problems in spacecraft dynamics, the development of formalisms for a computer-aided generation of equations of motion started. In this chapter we review the software for the generation of the equations of motion of multibody systems.

4.1 Introduction

Generating the equations of motion for complex multibody systems with a large number of degrees of freedom is difficult with paper and pencil. For this reason various computational methods for automatic equation generation have been developed.

The first formalisms have been intended to serve as the basis for a computer program for numerical generation, since the methods tend to be ill-suited for analytical investigations, e. g., ADAMS, Orleanda (1973), COMPAMM, Jiménez et. al. (1990), DADS, Haug (1984), DISCOS, Frisch (1990), MEDYNA, Duffek et. al. (1986), SIMPACK, Rulka (1990). Numerical programs are characterized by numerical digital computation, that is, starting with an initial condition, the behavior is analyzed by reiterated transformations of the basic model. Some of these programs are constructed by deriving once and for all, the equations of motion (dynamical and kinematical) of a very general system of interconnected bodies. In such derivations, the number of bodies in the system as well as the type of connection between adjacent bodies are treated as parameters; this leads to iteration loops and branching statements in the resulting program. This approach has obviously a number of drawbacks: it requires starting the computations all over again for each new set of input data and it does not permit all the insight provided by analytical equations, Fig. 4.1a. Moreover, for controller design numerically generated equations are of no help. Therefore, automatic derivation of symbolic equations of motion using a computer seems to be desirable even for simple models. Most recent multibody literature has dealt extensively with the formulation on an analytical basis, i. e., the equations produced are given as explicit formulas, e. g., AUTOLEV, Levinson and Kane (1990), MESA VERDE, Wittenburg and Wolz (1984), NEWEUL, Kreuzer (1979), SD-EXACT, and its new version SD-FAST, Rosenthal and Sherman (1983). The symbolic representation has the advantage that the equations have to be generated only once and the expressions have only to be evaluated during time-integration. The costs for the formulation of one symbolic set of equations is higher than once running through a numerical formalism. However, symbolic equations are an exact basis for the considered model and combined with a package for numerical analysis and simulation they result in ten or even more times faster computations than a purely numerical approch would require, Fig. 4.1b.

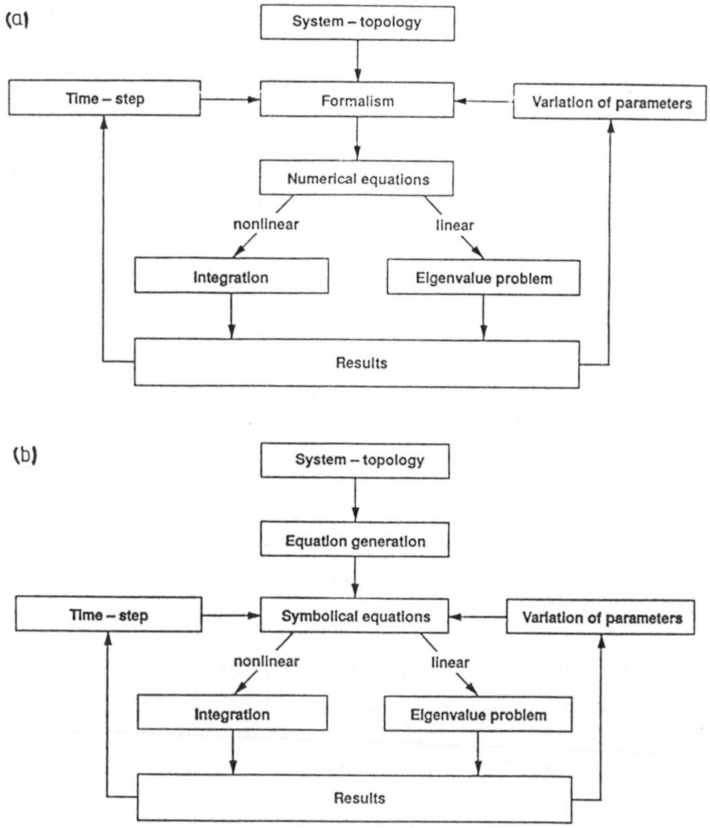

Figure 4.1: On the analysis of multibody systems: (a) numerical equation generation, (b) analytical equation generation

In a recently published multibody systems handbook a comparison of important multibody system dynamics software is presented, Schiehlen (1990).

4.2 General Discussion of Multibody Dynamics Software

The value of multibody system programs lies in the ease with which they can be used in the design process. They provide very comfortable simulation capabilities and, like other developments in computer aided design, their main goal is to increase productivity and to improve the quality and performance of the final design, Andrews (1983).

For a software package to prove as a useful and reliable tool, certain basic requirements have to be met; compare Kortüm and Schiehlen (1985):

- Handling: Software should be easily usable by engineers with minimal requests on input data; it should be as much as possible dialog-oriented with help-texts and logical tests for wrong input data. All input should be format-free. Semantic and syntactic checks on input should be performed.
- Documentation: Only software with full documentation, background material and user's manual is of use. Documentation must contain information about range of validity of the modeling, on approximations and assumptions.
- Reliability: A well-tested program uses proven and robust algorithms, so that correct results can be expected from a physically meaningful set of input data.
- Portability: To allow easy transfer, the software should be written in a high-level language for which compilers exist everywhere (i. e. today: FORTRAN). The computer-dependent (operation system) routines should be minimal and the producer should give an implementation guarantee.
- Integrity: For a certain area, such as vehicle dynamics, it is desirable to have one integrated software package available. This relieves the necessity of learning different concepts and languages for one type of problem, and also to buy tools from different sources (maintenance, interface problems etc.).
- Interoperability: Despite of the previous request, it should be possible to communicate with other main software tools (e. g. with Finite Element programs or with established simulation languages). Interoperability is mainly a question of the data management of the program.
- Efficiency: The efficiency of the software, i. e. the storage requirements (e. g. use of overlay or segmented structure) and the calculation-time are of interest. Both requirements, however, are not as stringent as in the past, because of the availability of virtual operating systems and the increase of computer speed.
- Maintenance: Software, which is in use will be improved and corrected permanently. The producer should offer services such as assistance, maintenance and updating.

If the above basic requirements on software are met, the user is interested further in the modeling options, the computational methods, the pre- and postprocessing capabilities and the presentation of the final results (graphics). These items are discussed in the following section.

4.3 Pre- and Postprocessing

Many powerful computer programs are nowadays available for deriving the equations of motion and to simulate the mathematical model of even complex mechanical systems efficiently. To improve the transformation of the real system into a multibody system, the interaction of multibody formalisms with CAD-systems has to be urgently pursued. According to the different multibody formalisms and computer codes, a new general description and organization of multibody system data has been introduced to fulfill the needs of a modular simulation system based upon numerical and symbolical formalisms, Otter et. al. (1993). With respect to different solid model construction methods, multibody system components are generated by extraction of an extended CAD data base, Daberkow and Kreuzer (1994).

In the last decade, CAD systems became crucial tools for designers in many areas of sciences, especially in mechanical design and engineering. In early stages of development, CAD systems were merely used to assist in the two-dimensional drawing process. Nowadays, these systems are widely embedded in the industrial design and construction process, while a throughout application of three-dimensional Solid Modeling CAD-systems is rare.

Encouraged by the increasing applications of multibody formalisms and CAD systems, the coupling of solid modelers with multibody simulation software became more and more attractive. For the numerical computer code ADAMS, Orleanda (1973), the graphic modeling tool P/Mechanism, included in the widespread pre- and postprocessing finite element design tool PATRAN (1987), supports a graphical modeling and postprocessing of simulation results. A further interface to this multibody formalism on Cartesian coordinates is provided by the CAD system ARIES (1989). All these systems have in common that they support the calculation of mass properties from the solid model designed in the CAD system. With additional mechanical properties, such as the locations of applied forces and torques and of joint attachment points, the solid models represent the rigid bodies of the multibody system.

The combination of computer aided design and simulation of multibody system's behavior in an automated design environment is desirable for various reasons. As multibody models became increasingly complex, the amount of engineering knowledge needed by a single designer is increasing to sometimes unmanagable amounts. This justifies a combined design and dynamic analysis approach based upon advanced CAD systems. In particular, a system dynamics investigation requires the basic parameters of mass, center of gravity, and moments of inertia without considering the geometry model, modeling method and solid model construction of the CAD system. A high degree of modularization demands an exchange of complete or single object data of a multibody system. Moreover, simulation results of a dynamic analysis have to be transferred to a CAD- or graphics-system without specific concern or restrictions on the geometry model representation. Furthermore, an efficient design process of mechanical parts requires a general interface to multibody computer codes, Fig. 4.2. This interface is claimed to serve as a compatible and comfortable post processor, taking the different algorithms and implementations of multibody dynamics formalisms into account. Moreover, the geometric information about attachment points of joints and force elements is expected to be evaluated in the design state. Commercially available multibody modeling software tools within CAD systems are mostly dedicated to a particular multibody dynamics computer code. Often, no options are supplied for a parametric multibody system description or the modeling is restricted to either robot, mechanism or vehicle dynamics. Consequently, the problem formulation and motivation leads to the following key concept:

- Comprise the necessary data describing a multibody model for the different multibody programs.
- Examine the different geometry models of CAD systems for solids and extract the relevant data for multibody systems.
- Define a geometry model for the representation of multibody elements.
- From these objectives, design data types and operations and construct a software interface for a system-independent modeling of multibody systems.

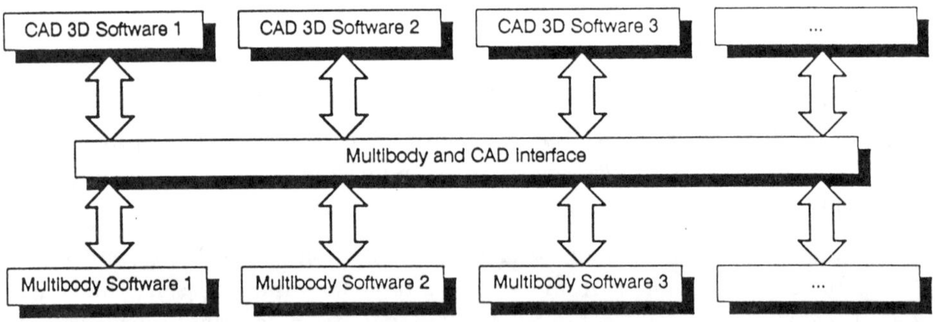

Figure 4.2: Automated model conversion

All these points have been realized in Daberkow (1993) where also details of the implementation are presented.

5 Special-Purpose Software for Generation of Symbolic Equations of Motion

In this chapter we describe the computer program NEWEUL that uses a special symbol manipulation technique together with Newton's and Euler's equations for automatically generating symbolic equations of motion for complex mechanical systems modeled as multibody systems. D'Alembert's or Jourdain's principle is used for reducing the system's order to the number of equations necessary. The equations so generated describe the motion, provide symbolic expressions for actuator forces and torques, and even allow the computation of joint-reaction forces. Furthermore, strength estimates necessary for the economic design of mechanical systems can also be ascertained. The generality of multi-purpose programs often prevents them from generating a computationally efficient simulation code. But with user defined simplification routines which make use of the special properties of a particular system, it is possible to arrive at efficient dynamic equations. The chapter closes with the application of the formalism to an industrial robot.

5.1 Introduction

This part adresses the general problem of developing special-purpose software to automate the derivation of explicit or symbolic equations of motion of multibody systems. Based upon the systematic computer oriented approach outlined so far the program package NEWEUL has been developed allowing to perform the necessary symbolic manipulation within the FORTRAN environment. It is not enough for the approach, however, to work well on simple problems that are limited in size and complexity but for large systems where hand generation becomes very difficult or nearly impossible. If nonlinear equations have to be derived, extensive expressions are involved. Without appropriate techniques, the derivation process can become time consuming and the equations generated can be very long as well as inefficient for further numerical computations. Thus, several problems must be taken care of that symbolic equation generation can become widely accepted and applied.

- The derivation of equations of motion must be made efficient to handle the large expressions especially in case of nonlinear problems.
- Methods such as simplification and compression must be employed to reduce inefficiencies that are usually associated with automatically generated equations.
- The system and its user interface must be designed for ease of use by engineers and scientists who have no extensive computer experience.

The system NEWEUL is a self-contained package written in FORTRAN. There are no restrictions concerning the computer. The design goals of NEWEUL as well as reasons for not using a commercially available symbol manipulation package are discussed in section 2. Section 3 describes symbolic computation in FORTRAN by means of index coding, Kreuzer (1979). Section 4 contains a brief discussion of the most advanced NEWEUL version and the data description. Aside from applications for multibody system dynamics, the formula manipulation scheme presented herein can also be used for symbolic matrix manipulation, Schmoll (1985).

5.2 Generation of Symbolic Equations

An analysis of the algorithmic description shows, that during the deriviation of the equations of motion only the operations

- addition,
- subtraction,
- multiplication, and
- differentiation

on vectors and matrices have to be performed. Moreover, simplification routines for trigonometrical expressions as well as routines for linearizing expressions are necessary. All these operations can be performed by means of general purpose symbol manipulation programs. Scientists who are concerned with solving problems in their own field would, naturally, prefer to employ an existing system to satisfy their needs. It would seem then that general-purpose systems might satisfy their needs. They are written to handle a large number of mathematical contingencies. Furthermore, these programs require too much computational overhead to be used as the implementation language for a symbolic multibody program. Taking into account these considerations it is appropriate to develop a special-purpose symbol manipulation program.

5.3 Special-Purpose Software NEWEUL

The software NEWEUL has been developed to accomplish the above tasks. The program is written in FORTRAN using a special coding scheme. Although every representation in a computer must be unambiguous, any given algebraic expression can be represented in a variety of equivalent ways. In order to represent an algebraic expression efficiently and compactly, we store the minimum information needed. This can be done as follows, Fig. 5.1. The variables and functions are stored in arrays for the names of variables and functions, respectively. Variables and functions are labelled by integers in the sequence they are encountered. The labels of variables are positive integers, the labels of functions are negative integers, indicating their positions in the arrays.

Arithmetic expression L*AL1*SIN(BE+2*GA)+BE*GA*COS(AL)

Array of variables	. L . AL .. AL1 ... BE ... GA ...
Index	. 2 . 4 .. 7 ... 14 ... 22 ...
Array of functions	SIN COS
Index	-1 -2

Coded expression

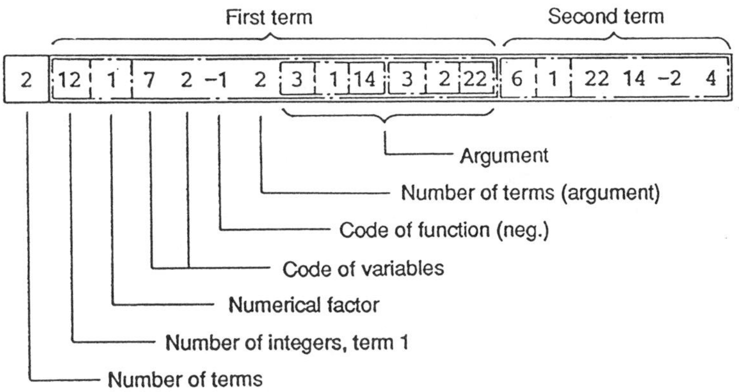

Figure 5.1: Representation of an expression by means of index coding

For storing and executing operations or expressions an array of integers, called task array, is used. This index coding is a simple and convenient means of representing the sequence of operations that apply to the variables and constants in an expression. We have to handle a sequential memory. This is done by pointers, representing the first code number of the expression in the task array. Vectors and matrices can be formed using symbolical expressions as matrix elements. Then a matrix is represented by an array of pointers, which are the indices of the elements in the task array. The program NEWEUL is based upon these basic ideas. The user of the system NEWEUL does, however, not have to know all the above details.

The ability to simplify expressions with functions applied either by the system or by the user, such as expansion, factorization, etc., is also desirable. This helps to reduce the problem caused by a natural phenomenon, "intermediate swell" or "expression pulsation", which may limit storage severely during processing.

Intermediate expression swell refers to the phenomena of intermediate computations producing expressions so large that their storage requirements exceed the capacities of the computer, even though the system being analyzed may be composed of only a modest number of elements. This problem is exacerbated by the fact that both the memory and processor time used by many important algebraic functions tend to grow exponentially while the size of the expressions operated on grows only linearly.

The programming of vector and matrix operations yields the complete program for the generation of the equations of motion of a multibody system where simplifications and intermediate variables are automatically introduced in a fashion that mimics the behavior of a human analyst.

NEWEUL is organized into different moduls. Each module performs a different subtask. The following features are available:

(a) The user interface module contains the menu-driven, interactive user environment for NEWEUL. The input handling features include: free format for all input with interactive prompting the correct input form; editing capabilities for correcting typing errors; the capability of saving all or part of the input for later use; and the flexibility of receiving input either interactively or from a text file.

(b) Symbolic operation as well as differentation of expressions, vectors, and matrices are involved.

(c) Based on easy-to-provide user input the geometry of the complete system is automatically generated.

(d) The equations produced are written into a file.

(e) Simplifications are implemented for automatic labelling of subexpressions to avoid re-computation in numerical procedures.

(f) Routines for taking advantages of symmetry and the automatic generation of FORTRAN functions are also included.

(g) Coding and de-coding of symbolic expressions on the basis of index-coding is done automatically. These routines include an expression parser, a code translator, and a decoding module.

(h) Test examples demonstrating specific features are provided also.

(i) An extensive documentation, a user's manual, and thoroughly explained examples are also included.

Porting NEWEUL on other machines is easy. Having an idea of the modular organization, we can proceed to examine the way NEWEUL works.

5.4 Fields of Application of the Software Package NEWEUL

The software package NEWEUL is designed to set up

- linear or
- nonlinear

equations of motion for multibody systems with

- holonomic and/or
- nonholonomic

constraints. Furthermore, symbolical expressions may be generated which enable the subpackage ZKSUB to numerically calculate the constraint forces occurring in joints/supports. The constraints may be scleronomic or rheonomic. The multibody system may have an open kinematic loop structure, a tree structure or a closed kinematic loop structure. General force laws may be included, too. Even if input and calculations are done thoroughly symbolically, constants may be automatically replaced by numerical values.

Systems with the above features are found e. g. in the following fields:

- vehicle dynamics,
- machine dynamics,
- robot dynamics,
- dynamics of mechanisms,
- satellite dynamics,
- biomechanics.

For further numerical analysis of the symbolical equations, e. g. calculation of eigenvalues or simulation, the results are always written in a format fully compatible to any FORTRAN program.

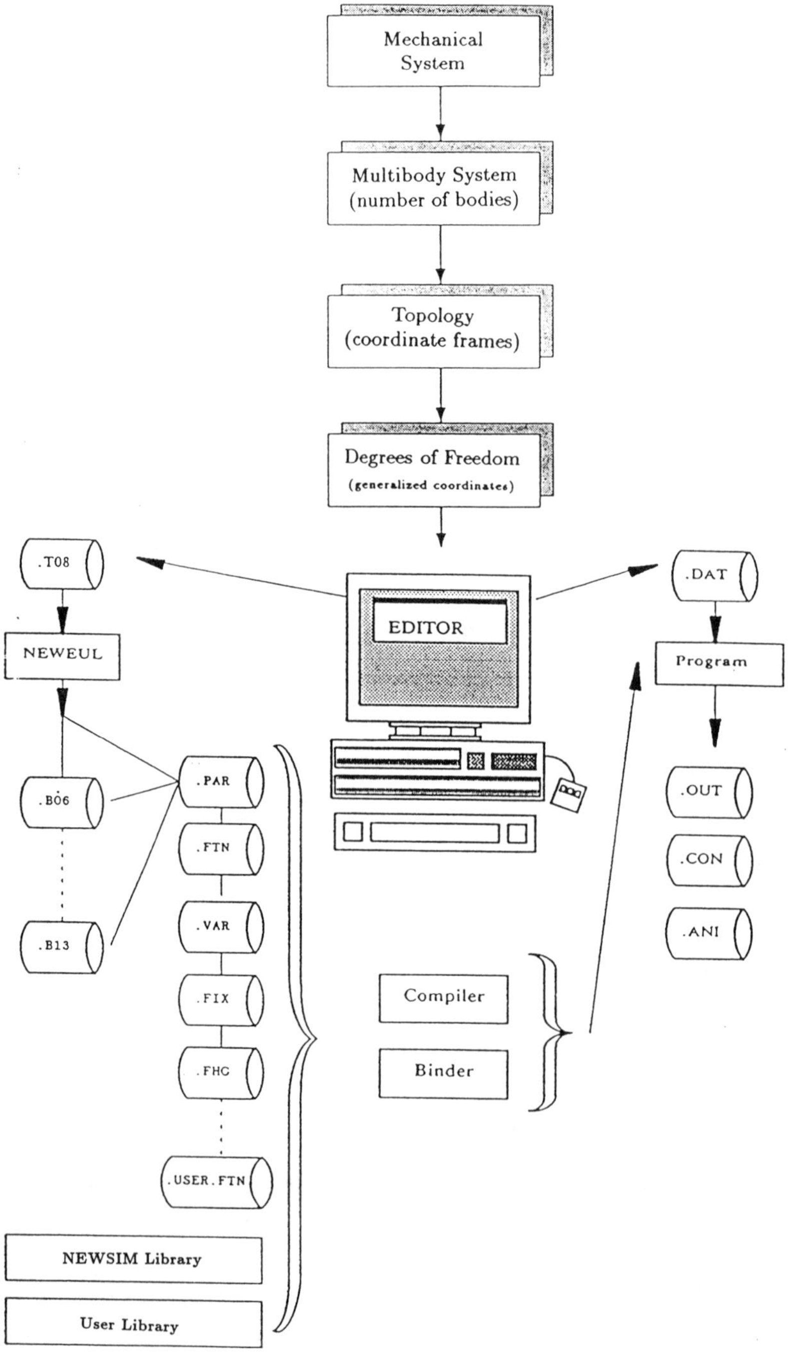

Figure 5.2: Procedure for mathematical modeling of multibody systems

5.4.1 Program Structure

The software package NEWEUL consists of many subroutines which are included in the main program NEWEUL. This program works as a batch-program, i. e. it reads input data from a file called TAPE08 and writes results on different files B06 ... B13, Fig. 5.2.

The file OUTPUT is displayed on the screen and contains all results produced in an easy-to-read format and for checking purposes also the complete set of input data.

For any system whose equations of motion are needed, one simply creates a partially prepared file using any text editor. This file contains the topology, geometry, and mass properties of the multibody system. The input file is prepared within an editor by means of filemasks. On the basis of these tools it is easy to summarize and formulate the complete set of input data needed.

The input file can be logically subdivided into six different sets which occur none, once or several times. These sets have to be ordered in a definite manner as shown in Fig. 5.3.

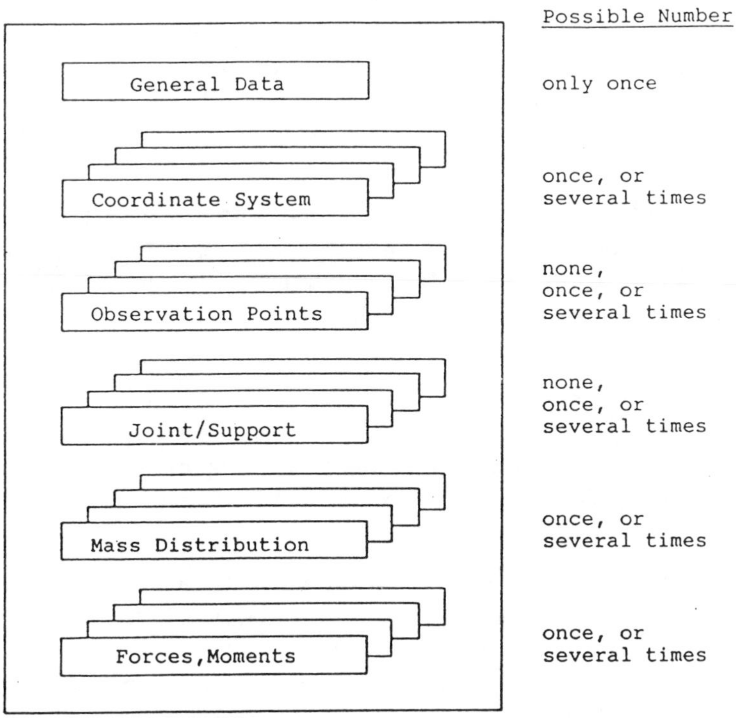

Figure 5.3: Structure of input file

5.4.2 Special Features of the Software Package

In the software package NEWEUL the following elements are available for modeling:

- rigid bodies,
- masspoints,
- joints/supports,
- nodal points,
- observation points.

Note that joints/supports and masspoints exclude each other and cannot be used together. Nodal points are always situated on a rigid body.

Reference Frames

The position of each element can be described with respect to an arbitrary reference frame. Then, the position of a rigid body K_i , $i = 1, \ldots, p$, is represented by the vector $\boldsymbol{r}_i$ from the origin O_R of the reference frame to its center of mass C_i, and the rotation matrix $\boldsymbol{S}_i$ describing the rotation of the body-fixed frame with respect to the reference frame.

The reference frame is allowed to be in arbitrary motion with respect to the inertial frame given by the translation vector $\boldsymbol{r}_R$ from the origin of the inertial frame to the origin of the reference frame and by the rotation matrix $\boldsymbol{S}_R$ describing the rotation of the reference frame relative to the inertial frame.

Relative Input

The postion of rigid bodies, masspoints and joints/supports can be described either with respect to a certain frame (whether inertial or reference) or relative to another element.

The reference element must be a rigid body or a joint/support. Its translation vector and rotation matrix have been stored by the program, i. e. the element must have been designated explicitly as reference element. This is the reason why the elements should properly numbered (especially rigid bodies and joints/supports). Relative input is always related to the origin of the frame of the reference element.

Nodal Points

Nodal points serve to describe the action of forces on bodies. A nodal point has to be assigned to a definite body. The situation depicted in Fig. 5.4 illustrates the application of nodal points. Force $\boldsymbol{F}$ acts on body K_i at nodal point A described by vector $\boldsymbol{r}$ from the body's center of mass C_i.

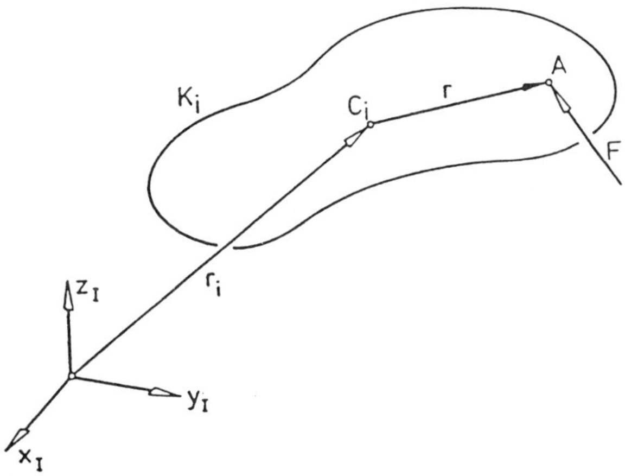

Figure 5.4: Example for application of nodal points

Using the nodal point approach, applied torques resulting from applied forces not acting at the center of mass of the considered body, are automatically included. Moreover, applied torques may by entered via nodal points as well.

Observation Points

Observation points allow to determine translational and rotational position, velocity and/or acceleration of arbitrary points of the multibody system with respect to any other point in an arbitrary reference frame. Observation points are very useful for the evaluation of applied forces, e. g. generated by tires.

5.4.3 NEWEUL Syntax Rules

The syntax for input is widely determined by the basic formula manipulation algorithm of index coding.

Allowed Characters

Allowed are capital letters, digits, the arithmetic operators plus mark, minus mark and multiplication mark and some special characters.

Variable Names

A name is a string of 1 – 6 alphanumeric characters beginning with a letter. As further restrictions must be observed:

- If time is used explicitly it must be denoted by the letter T, a variable always identified with time and therefore not allowed to identify other variables.
- Occasionally the program generates names by itself, namely in case of derivation of a position vector, differentation of auxiliary variables and calculation of compressed equations of motion. Names must be chosen so as to prevent ambiguities with these automatically generated names.

Constants

Constants can be integers or reals in a wide range depending on the computer in use. Constants beyond this range should not be used.

Operations, Brackets

Allowed operations are: +, −, *, **, i. e. addition, subtraction, multiplication and exponentiation. Long expressions are reduced according to the usual arithmetic and the usual bracket rules. Exponents must be positive integers.

Functions

Allowed functions are sine and cosine. Further functions can be represented by means of auxiliary variables and simplification equations. The argument may be a simple variable or an expression.

Error Codes

If an error occurs during the program execution and the program is able to detect it, NEWEUL indicates it by writing

```
* NEWEUL Fxx .
```

Here **xx** stands for the error code, the meaning of which is explained in the NEWEUL-Error Listing. After this message, calculation of the equations of motion by the program NEWEUL is immediately aborted.

5.4.4 Equations of Motion of an Industrial Robot

The mechanical modeling approach together with the geometric description developed above makes the immediate application of the NEWEUL program possible. First, let the physical and geometrical parameters (masses, inertias, mass center locations, connection joint locations) of the bodies of the system be read into the computer. All these data can be obtained easily from the system under consideration. To be more specific, the following data are required to run the NEWEUL program:

The global data for the system

$\boldsymbol{y}$	$f \times 1$-vector of generalized coordinates,
$\boldsymbol{z}$	$g \times 1$-vector of generalized velocities (optional),
$\boldsymbol{r}_R$	3×1-vector of the origin of the moving reference frame (optional),
$\boldsymbol{S}_R$	3×3-rotation matrix of the moving reference frame (optional),

and for each body i

$\boldsymbol{r}_i$	3×1-vector of the center of mass,
$\boldsymbol{S}_i$	3×3-rotation matrix of the body fixed frame,
m_i	mass,
$\boldsymbol{I}_i$	3×3-inertia tensor with respect to the center of mass,
$\boldsymbol{f}_i^a$	3×1-vector of applied forces,
$\boldsymbol{l}_i^a$	3×1-vector of applied torques.

With the entry program the determination of $\boldsymbol{r}_i$ and $\boldsymbol{S}_i$ using relative quantities is greatly simplified.

Finally, the symbolic and/or numeric equations of motion produced as FORTRAN-code are generated in batch mode. Linearization is accomplished by expanding equations of motion about the nominal operating point and retaining only the first order terms. These equations, together with the reaction equations, may then be used for further theoretical analysis, for model validation studies, and for numerical simulations. Thus, the user is free of the time-consuming and error-prone task of deriving equations of motion for each system he wishes to consider.

The complex dynamics of an industrial robot makes simulation and design of control laws very difficult. When neglecting elastic deformations of the links a manipulator may, approximately, be considered as a multibody system. If the individual motions of the active elements of a manipulator are neglected, then the number of rigid bodies equals the number of links. Considering these aspects the equations of motion of the manipulator schematically represented in Fig. 5.5 have been generated.

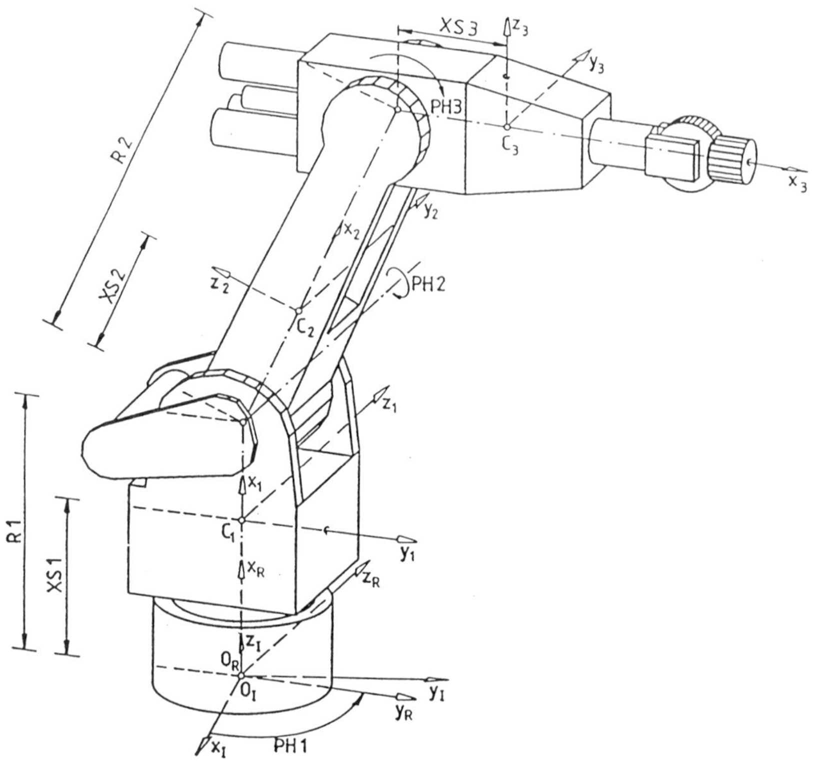

Figure 5.5: Robot manipulator

The system consists of three rigid bodies coupled to each other as indicated. In order to describe the unconstrained motion of the endpoint three degrees of freedom are necessary leading to the following coordinates:

$$\boldsymbol{y} = [\text{PH1} \quad \text{PH2} \quad \text{PH3}]^T \ . \tag{5.1}$$

The applied forces and torques for this system are found easily. They are due to the weight and to the actuators acting at the joints. From these data, supplemented by the masses and inertia tensors, the **NEWEUL** program generates the nonlinear equations of motion, Fig. 5.6. Introducing user-defined simplifications, the expressions might further be reduced. The input file is presented in the Appendix.

```
C>                          EQUATIONS OF MOTION
C>                          *******************
C>
C>
C>
C>    INERTIA MATRIX
C>
      RM(1,1)= A3*COS(PH3+PH2)**2 +
             + C3*SIN(PH3+PH2)**2 +
             + A2*COS(PH2)**2 +
             + C22*SIN(PH2)**2 +
             + R31**2*M3 +
             + A1
      RM(1,2)= 0.
      RM(1,3)= 0.

      RM(2,2)= M3*XS3**2*COS(PH3+PH2)**2 +
             + 2.*M3*XS3*R2*COS(PH3+PH2)*COS(PH2) +
             + M3*R2**2*COS(PH2)**2 +
             + R31**2*M3 +
             + B22 +
             + B3
      RM(2,3)= M3*XS3*R2*COS(PH3) +
             + B33

      RM(3,3)= B33
C>
C>
C>    VECTOR OF GENERALIZED GYROSCOPIC AND CENTRIFUGAL FORCES
C>
      K(1)= 2.*M3*XS3*R2*PH2P*PH1P*SIN(PH3+PH2)*COS(PH2) +
          + 2.*PH23P*M3*XS3*R2*PH1P*SIN(PH2)*COS(PH3+PH2) +
          + PH23P*C33A3*PH1P*SIN(2.*PH3+2.*PH2) +
          + M3*R2**2*PH2P*PH1P*SIN(2.*PH2) +
          + C22A2*PH2P*PH1P*SIN(2.*PH2)

      K(2)=-M3*XS3*R2*PH1P**2*SIN(PH3+PH2)*COS(PH2) -
          - M3*XS3*R2*PH1P**2*SIN(PH2)*COS(PH3+PH2) -
          - 0.5*C33A3*PH1P**2*SIN(2.*PH3+2.*PH2) -
          - 0.5*M3*R2**2*PH1P**2*SIN(2.*PH2) -
          - 0.5*C22A2*PH1P**2*SIN(2.*PH2) -
          - 2.*M3*XS3*R2*PH3P*PH2P*SIN(PH3) -
          - M3*XS3*R2*PH3P**2*SIN(PH3)

      K(3)=-M3*XS3*R2*PH1P**2*SIN(PH2)*COS(PH3+PH2) -
          - 0.5*C33A3*PH1P**2*SIN(2.*PH3+2.*PH2) +
          + M3*XS3*R2*PH2P**2*SIN(PH3)
C>
C>
C>    VECTOR OF GENERALIZED APPLIED FORCES
C>
      Q(1)= MM1

      Q(2)= GE*XS2*M2*SIN(PH2) +
          + GE*R31*M3 +
          + MM2

      Q(3)= GE*M3*XS3*SIN(PH3+PH2) +
          + MM3
```

Figure 5.6: Equations of motion

References

Andrews, G. C. (1983), A Brief Survey of "Self-Formulating" Simulation Programs for Multibody Dynamic Systems. In: Proc. IASTED Intern. Symp. Simulation and Modeling.

Bremer, H. (1988), Dynamik und Regelung mechanischer Systeme. Stuttgart: Teubner.

Buchberger, B., Collins, G.E., and Loos, R. (eds.) (1982), Computer Algebra — Symbolic and Algebraic Computation. Wien, New York: Springer-Verlag.

Daberkow, A. (1993), Zur CAD-gestützten Modellierung von Mehrkörpersystemen. Fortschritt-Berichte der VDI-Zeitschriften, Reihe 20, Nr. 80, Düsseldorf, VDI-Verlag.

Daberkow, A., and Kreuzer, E. (1994), An Integrated Approach for Computer Aided Design in Multibody System Dynamics. To appear.

Davis, M.S. (1978), Analytical Mathematics on Computers. Applied Mechanics Reviews, 31, 1 – 9.

Duffek, W., Führer, C., Schwartz, W., and Wallrapp, O. (1986), Analysis and Simulation of Rail and Road Vehicles with the program MEDYNA. In: Proc. 9th IAVSD-Symposium on the Dynamics of Vehicles on Roads and on Tracks, Nordström, O. (ed.), Lisse: Swets and Zeitlinger, 71 – 85.

Frisch, H. P. (1990), NBOD & DISCOS — Dynamic Interaction Simulation of Controls and Structure. In: Multibody Systems Handbook, Schiehlen, W. (ed.), Berlin/...: Springer-Verlag, 145 –160.

Fitch, J. (1993), Mathematics goes automatic. Physics World, 48 – 52.

Grossmann, R. (ed.) (1989), Symbolic computation: application to scientific computing. Philadelphia: Soc. for Ind. and Appl. Math. (SIAM).

Haug, E. J. (1984), Elements and Methods of Computational Dynamics. In: Computer Aided Analysis and Optimization of Mechanical System Dynamics, Haug, E. J. (ed.), Berlin/...: Springer-Verlag, 3 – 38.

Jiménez, J. M., Avello, A., Garcia-Alonso, A., and Garcia de Jalón, J. (1990), COMPAMM — A Simple and Efficient Code for Kinematic and Dynamic Numerical Simulation of 3-D Multibody Systems with Realistic Graphics. In: Multibody Systems Handbook, Schiehlen, W. (ed.), Berlin/...: Springer-Verlag, 285 – 304.

Kahrimanian, H.G. (1954), Analytical Differentiation by a Digital Computer. Sympos. on Automatic Programm. for Digital Comput., Off. of Naval Res., Dept. of the Navy, 6 – 14.

Kortüm, W. and Schiehlen, W. (1985), General Purpose Vehicle System Dynamics Software Based on Multibody Formalisms. Vehicle System Dynamics, 14, 229 – 263.

Kreuzer, E. (1979), Symbolische Berechnung der Bewegungsgleichungen von Mehrkörpersystemen. Fortschritt-Berichte der VDI-Zeitschriften, Reihe 11, Nr. 32, Düsseldorf: VDI-Verlag.

Kreuzer, E. and Leister, G. (1991), Programmpaket NEWEUL'90. Stuttgart: Universität, Inst. B für Mechanik, Anleitung AN-23.

Kreuzer, E. and Schiehlen, W. (1990), NEWEUL-Software for the Generation of Symbolical Equations of Motion. In: Multibody Systems Handbook, Schiehlen, W. (ed.), Berlin/...: Springer-Verlag, 181 – 202.

Leister, G. (1992), Beschreibung und Simulation von Mehrkörpersystemen mit geschlossenen kinematischen Schleifen. Fortschritt-Berichte der VDI-Zeitschriften, Reihe 11, Nr. 167, Düsseldorf, VDI-Verlag.

Leister, G. and Bestle, D. (1992), Symbolic-numerical Solution of Multibody Systems with Closed Loops. International Journal of Vehicle Design.

Levinson, D. A. and Kane, T. R. (1990), AUTOLEV - A New Approach to Multibody Dynamics. In: Multibody Systems Handbook, Schiehlen, W. (ed.), Berlin / ...: Springer-Verlag, 81 – 102.

Loos, R. (1982), Introduction. In: Computer Algebra — Symbolic and Algebraic Computation. Buchberger, B., Collins, G.E., and Loos, R. (eds.), Computing Supplementum 4. Wien, New York: Springer-Verlag.

Nolan, J. (1953), Analytical Differentiation on a Digital Computer. Cambridge, Mass.: Inst. of Technol. (MIT), M.A. Thesis.

Noor, A.K. and Andersen, C.M. (1979), Computerized Symbolic Manipulation in Structural Mechanics — Progress and Potential. Computers and Structures, 18, 95 – 118.

Orleanda, N. (1973), Node-Analogous Sparsity-Oriented Methods for Simulation of Mechanical Systems. Ph. D. dissertation, University of Michigan.

Otter, M., Hooke, M., Daberkow, A., and Leister, G. (1993), An Object-Oriented Data Modul for Multibody Systems. In: Advanced Multibody System Dynamics

— Simulation and Software Tools, Schiehlen, W. (ed.). Dordrecht/...: Kluwer Academic Publishers, 19 – 48.

Parker, T.S. and Chua, L.O. (1989), Practical Numerical Algorithms for Chaotic Systems. New York/...: Springer-Verlag.

Pavelle, R., Rothstein, M., and Fitch, J. (1981), Computer Algebra. Scientific American, 136 – 152.

Rand, R.H. and Armbruster, D. (1987), Perturbation Methods, Bifurcation Theory and Computer Algebra. New York/...: Springer-Verlag.

Rauh, J. (1987), Ein Beitrag zur Modellierung elastischer Balkensysteme. Fortschritt-Berichte der VDI-Zeitschriften, Reihe 18, Nr. 37, Düsseldorf, VDI-Verlag.

Rosenthal, D.E. and Sherman, M.A. (1986), High performance multibody simulation via symbolic equation manipulation and Kane's method. J. of Astronautical Sciences, 34, 223 – 239.

Rulka, W. (1990), SIMPACK — A Computer Program for Simulation of Large-Motion Multibody Systems. In: Multibody Systems Handbook, Schiehlen, W. (ed.), Berlin / ...: Springer-Verlag, 265 – 284.

Schiehlen, W. (ed.) (1990), Multibody Systems Handbook. Berlin/...: Springer-Verlag.

Schirm, W. (1993), Symbolisch-numerische Behandlung von kinematischen Schleifen in Mehrkörpersystemen. Fortschritt-Berichte der VDI-Zeitschriften, Reihe 11, Nr. 198, Düsseldorf, VDI-Verlag.

Schmoll, K.-P. (1988), Modularer Aufbau von Mehrkörper-Systemen unter Verwendung der Relativkinematik, Fortschritt-Berichte der VDI-Zeitschriften, Reihe 18, Nr. 57, Düsseldorf: VDI-Verlag.

Schramm, D. (1986), Ein Beitrag zur Dynamik reibungsbehafteter Mehrkörpersysteme. Fortschritt-Berichte der VDI-Zeitschriften, Reihe 18, Nr. 32, Düsseldorf, VDI-Verlag.

Sunada, W.H. and Dubowsky, S. (1982), On the Dynamic Analysis and Behavior of Industrial Robotic Manipulators with Elastic Members. ASME Journal of Mechanical Design 1–10.

Truckenbrodt, A. (1980): Bewegungsverhalten und Regelung hybrider Mehrkörpersysteme mit Anwendung auf Industrieroboter. Fortschritt-Berichte der VDI-Zeitschriften, Reihe 8, Nr. 33, Düsseldorf: VDI-Verlag.

Wittenburg, J. and Wolz, U. (1985), MESA VERDE — Ein Computerprogramm zur Simulation der nichtlinearen Dynamik von Vielkörpersystemen. Robotersysteme, 1, 7 – 18.

Wolfram, S. (1988), Mathematica — A System for Doing Mathematics by Computer. Redwood City/...: Addison-Wesley Publ. Comp.

Zeis, E. and Cefola, P. (1980), Computerized Algebraic Utilities for the Construction of Nonsingular Satellite Theories. J. Guidance and Control, 3, 48 – 54.

N.N. (1989), ARIES Conceptstation Software Simulation Mechanism Reference. Aries Technology Inc., Lowell, M.A.

N.N. (1987), PATRAN Mechanical System Simulation, P/Mechanism User's Guide. PDA Engineering, Costa Mesa, CA.

Appendix

Record of the computations of the equations of motion by the program NEWEUL

```
C>  Eingabefile fuer NEWEUL'87 erstellt mit Editor
C>
C>    Knickarm-Roboter NOKIA 06
C>
C>  $$$$$$$$$$$$$$$$$$$$$$$$$$$$$$$$$$$$$$$$$$$$$$$$$$$$$$$$$$$$$$$$$
C>
      0   Steuerparameter: 0=holonom, 1=nichtholonom
      0   Komprimierungart(0-5): 0=vollsymbolisch
      4   Zahl der KOORDINATENSYSTEME
      3   Zahl der (Lage-)Freiheitsgrade
C>    0   Zahl der (Geschwindigkeits-)Freiheitsgrade
      0   Zahl der Hilfsvarablen
      0   Zahl der linearisierbaren Groessen
      0   Zahl der numerischen Groessen
      10   Zahl der Vereinfachungen
      0   Zahl der Substitutionsvariable
/IAFORM=3/
C> /MAXIF=500000/MAXRF=900/MAXIZF=550/INULL=0/LKINEM=1/
C>
C>
C>    LAGEVEKTOR
C>   ************
C>
   Y:PH1
   Y:PH2
   Y:PH3
C>
C>
      2   Automatische Ableitung (1: 1/2; 2: P/PP; 0: keine)
C>
C>
C>    Vereinfachungen
C>   ************
   VFV:PH2P+PH3P
   VFE:PH23P
C>
   VFV:SIN(PH2)*COS(PH3)+SIN(PH3)*COS(PH2)
   VFE:SIN(PH3+PH2)
C>
```

```
   VFV:COS(PH3)*COS(PH2)-SIN(PH3)*SIN(PH2)
   VFE:COS(PH3+PH2)
C>
   VFV:XS2**2*M2+B2
   VFE:B22
C>
   VFV:M3*XS3**2*SIN(PH3+PH2)**2+B3
   VFE:B33
C>
   VFV:XS2**2*M2+C2
   VFE:C22
C>
   VFV:XS3**2*M3+C3
   VFE:C33
C>
   VFV:C22-A2
   VFE:C22A2
C>
   VFV:C33-A3
   VFE:C33A3
C>
   VFV:XS3*SIN(PH3+PH2)+R2*SIN(PH2)
   VFE:R31
C>
C>
C>
C>    KOORDINATENSYSTEM  Nr.  1
C>    ***************************
C>
KOSART:   R                    Klassifizierung des Koordinatensystems
KOSYNA:   REFSYS               Namen des Koordinatensystems
KOSYNA:   I                    Namen des Bezugssystems
C>
C>   ****  Rotatorischer Teil  ****
      1  Zahl der Teildrehungen
C>
C>   Teildrehung Nr.
     4   Art der Drehung (-3-+5)
C>
C>   Drehungsmatrix
   S(1,1)=0.
   S(1,2)=COS(PH1)
   S(1,3)=-SIN(PH1)
   S(2,1)=0.
   S(2,2)=SIN(PH1)
```

```
   S(2,3)=COS(PH1)
   S(3,1)=1.
   S(3,2)=0.
   S(3,3)=0.
C>
C>   ****  Translatorischer Teil   ****
      1   Zahl der Teilvektoren
C>
C>   Teilvektor Nr.
KOSYNA:  I          System fuer Teilvektor
C>
C>   Teilvektor
   R(1)=0.
   R(2)=0.
   R(3)=0.
C>
C>
C>    KOORDINATENSYSTEM  Nr.  2
C>   ***************************
C>
KOSART:   S                  Klassifizierung des Koordinatensystems
KOSYNA:   ARM1               Namen des Koordinatensystems
KOSYNA:   REFSYS             Namen des Bezugssystems
C>
C>   ****  Rotatorischer Teil  ****
      0  Zahl der Teildrehungen
C>
C>
C>   ****  Translatorischer Teil   ****
      1   Zahl der Teilvektoren
C>
C>   Teilvektor Nr.
KOSYNA:  REFSYS          System fuer Teilvektor
C>
C>   Teilvektor
   R(1)=XS1
   R(2)=0.
   R(3)=0.
C>
C>
C>
C>    KOORDINATENSYSTEM  Nr.  3
C>   ***************************
C>
KOSART:   S                  Klassifizierung des Koordinatensystems
```

```
KOSYNA:   ARM2                  Namen des Koordinatensystems
KOSYNA:   REFSYS                Namen des Bezugssystems
C>
C>    ****  Rotatorischer Teil  ****
       1   Zahl der Teildrehungen
C>
C>    Teildrehung Nr.
      4   Art der Drehung (-3-+5)
C>
C>    Drehungsmatrix
   S(1,1)=COS(PH2)
   S(1,2)=0.
   S(1,3)=SIN(PH2)
   S(2,1)=SIN(PH2)
   S(2,2)=0.
   S(2,3)=-COS(PH2)
   S(3,1)=0.
   S(3,2)=1.
   S(3,3)=0.
C>
C>    ****  Translatorischer Teil   ****
       1   Zahl der Teilvektoren
C>
C>    Teilvektor Nr.
KOSYNA:  REFSYS                 System fuer Teilvektor
C>
C>    Teilvektor
   R(1)=XS2*COS(PH2)+R1
   R(2)=XS2*SIN(PH2)
   R(3)=0.
C>
C>
C>     KOORDINATENSYSTEM  Nr.  4
C>    ***************************
C>
KOSART:   S                     Klassifizierung des Koordinatensystems
KOSYNA:   ARM3                  Namen des Koordinatensystems
KOSYNA:   REFSYS                Namen des Bezugssystem
C>
C>    ****  Rotatorischer Teil  ****
       1   Zahl der Teildrehungen
C>
C>    Teildrehung Nr.
      4   Art der Drehung (-3-+5)
C>
```

```
C>    Drehung
    S(1,1)=COS(PH3+PH2)
    S(1,2)=0.
    S(1,3)=SIN(PH3+PH2)
    S(2,1)=SIN(PH3+PH2)
    S(2,2)=0.
    S(2,3)=-COS(PH3+PH2)
    S(3,1)=0.
    S(3,2)=1.
    S(3,3)=0.
C>
C>    ****  Translatorischer Teil   ****
       1   Zahl der Teilvektoren
C>
C>    Teilvektor Nr.
KOSYNA:   REFSYS            System fuer Teilvektor
C>
C>    Teilvektor
    R(1)=XS3*COS(PH3+PH2)+R2*COS(PH2)+R1
    R(2)=XS3*SIN(PH3+PH2)+R2*SIN(PH2)
    R(3)=0.
C>
C>
C>
C>               Massengeometrische Daten
C>              **************************
C>
C>    Massengeometrische Daten fuer
KOSYNA:   ARM1    Namen der Koordinatensystems
C>
C>    Masse
  MASS=   M1
C>
C>    Traegheitstensor
KOSYNA:   ARM1    System fuer Angabe der Traegheitstensors
C>
  I(1,1)= A1
  I(2,1)= 0.
  I(2,2)= B1
  I(3,1)= 0.
  I(3,2)= 0.
  I(3,3)= C1
C>
C>
C>    Massengeometrische Daten fuer
```

```
KOSYNA:   ARM2    Namen der Koordinatensystems
C>
C>   Masse
  MASS=   M2
C>
C>   Traegheitstensor
KOSYNA:   ARM2    System fuer Angabe der Traegheitstensors
C>
  I(1,1)= A2
  I(2,1)= 0.
  I(2,2)= B2
  I(3,1)= 0.
  I(3,2)= 0.
  I(3,3)= C2
C>
C>
C>   Massengeometrische Daten fuer
KOSYNA:   ARM3    Namen der Koordinatensystems
C>
C>   Masse
  MASS=   M3
C>
C>   Traegheitstensor
KOSYNA:   ARM3    System fuer Angabe der Traegheitstensors
C>
  I(1,1)= A3
  I(2,1)= 0.
  I(2,2)= B3
  I(3,1)= 0.
  I(3,2)= 0.
  I(3,3)= C3
C>
C>
C>              Eingepraefte Kraefte und Momente
C>            ************************************
C>
C>   Gewichtskraft
FLEART:    GK     Art der Kraft/des Moments (GK,AK,AM,IK,IM,$END)
C>
C>   Namen oder Konstante der Gravitation
GRAVI=  GE
C>
     -3        Richtung der Gravitation (+-1,2,3: +-X-,Y-,Z-Achse, 0: Eingabe)
C>
C>   **  Aeussere/innere Kraft // Aeusseres/inneres Moment   **
```

```
FLEART:   AM      Art der Kraft/des Moments (GK,AK,AM,IK,IM,$END)
C>
C>   System(e) auf das(die) die Kraft/das Moment wirkt(en)
KOSYNA:   ARM1           1.System (Wirkung positiv)
C>
C>   System, in dem die Kraft/das Moment eingegeben wird
KOSYNA:   ARM1
C>
C>   Kraft/Moment eingeben
   FLE(1)=MM1
   FLE(2)=0.
   FLE(3)=0.
C>
C>
C>   **  Aeussere/innere Kraft // Aeusseres/inneres Moment  **
FLEART:   IM      Art der Kraft/des Moments (GK,AK,AM,IK,IM,$END)
C>
C>   System(e) auf das(die) die Kraft/das Moment wirkt(en)
KOSYNA:   ARM2           1.System (Wirkung positiv)
KOSYNA:   ARM1           2.System (Wirkung negativ)
C>
C>   System, in dem die Kraft/das Moment eingegeben wird
KOSYNA:   ARM2
C>
C>   Kraft/Moment eingeben
   FLE(1)=0.
   FLE(2)=MM2
   FLE(3)=0.
C>
C>
C>   **  Aeussere/innere Kraft // Aeusseres/inneres Moment  **
FLEART:   IM      Art der Kraft/des Moments (GK,AK,AM,IK,IM,$END)
C>
C>   System(e) auf das(die) die Kraft/das Moment wirkt(en)
KOSYNA:   ARM3               1.System (Wirkung positiv)
KOSYNA:   ARM2               2.System (Wirkung negativ)
C>
C>   System, in dem die Kraft/das Moment eingegeben wird
KOSYNA:   ARM3
C>
C>   Kraft/Moment eingeben
   FLE(1)=0.
   FLE(2)=MM3
   FLE(3)=0.
C>
```

```
FLEART:   $END    Art der Kraft/des Moments (GK,AK,AM,IK,IM,$END)
```

COMPUTER ALGEBRA SOFTWARE FOR SCIENTIFIC APPLICATIONS

F. Schwarz
GMD, Sankt Augustin, Germany

1 Introduction

Computer Algebra may be pursued within a wide range of sophistication. At the lowest level there is the casual user who applies some of the system facilities e.g. to evaluate an integral, to sum a series or to find the solution of a differential equation. For this kind of applications it is usually sufficient to read through the manual which is provided by the system developer. After a while it is realized that the machine might perform additional tasks for which there are no system functions available. So the user starts writing his own special purpose package. As the time goes on additional facilities are added which means that the size of the software keeps growing. That works fine for a while until a critical size of about one thousand lines of code is reached. It suddenly turns out that the time which has to be spent for debugging or adding a new feature has increased to an unacceptable amount. Even worse, that process resembles more and more a vicious circle where a simple change in one part of the program causes unexpected effects in completely different regions and a superficial repair often makes things even worse. While spending most of the time for implementing new algorithms for new tasks, somehow the survey over the organization of the package got lost. At that point the astute user will realize that he has reached a threshold where qualitatively new tools for further proceeding are needed. To provide these tools is the realm of *software engineering.*

It is the purpose of this article to discuss the problems that arise in connection with the development of computer algebra software in the broader context of software engineering and object-oriented design. In the remaining part of this introduction the most important general purpose computer algebra systems are shortly presented. In the subsequent section 2 a review of some concepts of software engineering with particular emphasis on object-oriented design is given. It is accompanied by a fairly extensive list of literature on this subject. In sections 3 and 4 typical features of two utterly different computer algebra systems, i. e. *REDUCE* and *Axiom*, are discussed. These sections are not meant as a substitute for the literature on the respective system that is provided by its developers. Rather they should be considered as a complementary reading material describing some of its implementation peculiarities in the context of section 2. Section 5 introduces the concept of *system independent software* that is based on a set of standardized types and a high level language, providing the interface between mathematical problems and the machine. It is illustrated by several examples that are worked out in detail, e. g. the symmetry package *SPDE* and the package *FINDS* for determining first integrals of dynamical systems. In the final section 6 a summary of the topics covered in this article and an outlook for future work are given.

A first overview of computer algebra systems. Some work on analytical differentiation in the early fifties is often considered to be the first attempt to use a digital computer for symbolic or algebraic manipulations. Now, about 40 years later more than one hundred systems exist for doing some form of computer algebra. An important distinctive feature for its classification is the area of application which the originators had in mind during its development. In the first place this entails the classification into *special purpose systems* and *general purpose systems.* Examples of the former species are systems like *CAYLEY* for computations with finite groups, *Macauley* for computing with polynomial ideals or *NEWEUL* [1] for the automatic generation of the equations of motion in mechanical engineering. These systems are usually more efficient and better adapted for its area of specialization than a general purpose system.

On the other hand, general purpose systems provide facilities for doing basic symbolic computations in almost all areas of mathematics. The range of these facilities grows larger in the course of time. In order to qualify for being a general purpose system, at present the following features are considered necessary.

- Operations on integers and rational numbers with unlimited accuracy; computations in finite fields and facilities to compute with algebraic numbers.
- Operations on polynomials and rational functions in one or more variables; basic arithmetic on these objects including gcd calculations and factorization over various coefficient domains.
- Gröbner base calculations and facilities for solving systems of algebraic equations pursuing various strategies for the solution.
- Formal differentiation and integration of rational, elementary and algebraic functions. At least some rudiments of an automatic differential equation solver.
- More special features like calculations of limits, basic tensor calculus and a symbolic-numeric interface, e. g. automatic generation of FORTRAN or C code.
- System features like an interpreter for interactive use, graphics tools, special editors and debuggers.

In Table 1 the general purpose computer algebra systems with the largest distribution are listed. *Derive* is a small system mainly for educational purposes. It is a perfect entrance in the area of symbolic computation and has a wide distribution now even on high schools. *REDUCE* is one of the oldest systems. It originated in the early sixties from high energy physics applications. Due to its availability on virtually all platforms (PC, Unix and AIX workstations etc.) it has a wide distribution. It is a completely open system. *MAPLE* is primarily designed for interactive use. It provides a large functionality and is suited for educational purposes as well. *Axiom* (formerly Scratchpad II) of the computer algebra group at IBM in Yorktown Heights is the first 3rd generation system which applies most principles

Name	Language	Principal Originators
MACSYMA	LISP	Mathlab Group, MIT
MAPLE	C	K. Geddes, G. Gonnet
Mathematica	C	S. Wolfram
Derive	LISP	D. Stoutemyer
REDUCE	RLISP	A. C. Hearn
Axiom	LISP	D. Yun, R. D. Jenks

Table 1: The most important general purpose computer algebra systems.

of modern software engineering. In this respect it is significantly different from all the other systems.

The following question is frequently raised when discussing the differences and the common features of computer algebra systems: What is the best system? A well-founded answer may only be given if it is complemented by the additional question: For what purpose? The answer depends only to a minor extent on the specific field of application. More important is this: How many lines of code does a user write by himself in order to obtain the functionality that he needs in his applications? There are several extremes possible. On the one hand, a user may simply apply some of the functions provided by the system interactively, very much like a numerical calculator is used. On the other hand, a user may decide to develop large software packages for special areas of application, possibly comprising several tens of thousands lines of code. The problems that have to be dealt with in these large-scale applications are completely different from those of an ad-hoc user of some system functions. Questions like how does a system support the extension by its design, does it provide tools for testing and debugging etc. become crucial. They are the main subject of this article.

Frequently computer algebra systems are distinguished according to whether they are *symbol-oriented* or *math-oriented.* In a symbol-oriented system, mathematical expressions are usually represented by a single or a few tree-like datastructures. Computation is performed primarily by tree traversal. There is a close mapping between what the user submits from his keyboard and the internal representation. The output is essentially an infix printing of the internal data structures. In a math-oriented system there may be little correspondence between the data structures applied internally and what the user submits or sees as output. Any mathematical object has an internal representation that is well adapted to the operations which are permitted on it. In the terminology to be introduced in the subsequent sections, mathematical objects have a type and a set of allowed operations on them determined by this type. It turns out that *Axiom* with its object-oriented design is a typical representative of a math-oriented system, whereas all the other systems are more or less symbol oriented.

There are still further aspects for evaluating computer algebra systems which are particularly connected to the ever-decreasing price-performance ratio of computer hardware. One may ask for example to what extent a system supports parallelism in order to increase the size of problems that may be treated, or to what extent is a system free of bugs, i. e. how many problems may be run in the average before an error occurs?

2 Principles of Software Engineering

This section contains a short review of those concepts from software engineering that are relevant for any serious user of a computer algebra system. It turns out that many apsects that arise during the design and the implementation of computer algebra software may be better understood if they are considered in this more general context.

2.1 Software Development

> **Abstraction.** *The principle of ignoring those aspects of a subject that are not relevant to the current purpose in order to concentrate more fully on those that are.*
>
> Oxford Dictionary of Computing, 1986

The principal goal of software development may bluntly be expressed as follows: Design and implement a piece of software that solves the problems from some application field efficiently and safely. According to Meyer [13], this goal may be reduced to the following *software quality measures.*

- *Correctness* is the ability of software products to exactly perform their tasks as defined by requirements and specification. This is clearly the prime quality because if a system does not do what it is supposed to do, then everything else matters little.
- *Robustness* or *reliablity* is the ability of software systems to function even in abnormal conditions, i. e. in those cases that are not explicitly covered by the specification. In other words, it must be difficult to break your software.
- *Extendibility* is the ease with which software products may be adapted to changes of specification. This is important because the requirements will certainly change and most likely increase if a piece of software is applied by many users. This is a typical issue that concerns the *programming-in-the-large.*
- *Efficiency* is the good use of hardware resources such as processors, internal and external memory etc. It will become apparent later on that there is a close connection between the overall efficiency of a piece of software and its internal design, e. g. how well datastructures conform to the objects of a certain application area.
- *Portability* is the ease with which software products may be transferred to various hardware and software environments, reaching for example from a MS-DOS environment on a notebook like PC to a large UNIX based workstation.

After the goals of software development have been clearly identified, the question as to how to arrive at these goals has to be considered. A little consideration will lead to the conclusion that the ultimate reason for most of the problems which occur during the development and the maintenance of large pieces of software is the fact that too many lines of code have to be considered at a single time, i.e. the *complexity* has become too large. After this has been realized, the two most important principles of software development follow rather naturally.

- Decompose large pieces of software into smaller ones through *modularity*.
- Limit the growth of software through *reusability*.

As a consequence of the first principle, to enhance the simplicity of software it has to be *structured* such that individual pieces can be considered, implemented, fixed and changed with minimal consideration or effect on the other pieces. The term *module* is used to denote such a piece of software having a name which may be referred to by other parts of the program. The goal is to create modules that are themselves small programs that interact with one another in simple, well defined ways.

It turns out that *data abstraction* is the unique principle to apply in order to obtain an appropriate decomposition of large computer programs into modules. For an excellent introduction into this subject the book by Liskov and Guttag [14] is strongly recommended. *Abstract data types* are defined in terms of the operations which the system provides for manipulating these objects. This set includes operations to create objects, to obtain information from them and to modify them. The internal details of the representation are not relevant and need not even be known at that moment. This principle of *information hiding* leads to a high degree of *locality*. The *specification* defines the outside behavior of an abstract data type. The language which is used for it is adjusted to the respective level of abstraction. Accordingly a complex problem is not regarded immediately in terms of computer instructions but rather in terms of entities natural to the problem itself in some suitable abstraction. The operations on this uppermost level are then subjected to decomposition leading to the next lower level of abstraction using its own data types and language. This process of *stepwise refinement* continues until a level is reached that can be understood by a computer, i.e. a high-level programming language [18]. On each level only certain aspects of the full problem are considered leading in this way to the structuring which is aimed for. The relation between the various levels of computer algebra software is illustrated in Figure 1.

Abstraction level	Datatype	Specification
User interface	Mathematical expressions	Plain English
⋮	⋮	⋮
i-th level	i-th Abstract data type	i-th Pseudocode
⋮	⋮	⋮
System	System types	System code

Figure 1: The location of user defined abstract data types between the description in mathematical terms and the implementation level is shown.

In order to benefit most from a modular software design, a few principles should be observed. The modules must *correspond to syntactic units* in the underlying language. This precludes for example a module structure defining a module X as extending from lines M to N of the implementation code. Every module should *communicate with as few others as possible* because the effect of a change or of an error in a certain module may propagate to a large number of other modules through these connections. The *principle of weak coupling* requires that if any two modules communicate at all, they should exchange as little information as possible, and this must be obvious from the text of the modules involved.

Reusability of a small number of invariant components for the construction of complex systems has been a key principle in mechanical or electrical engineering for a long time, e. g. when constructing a car, a building or an electronic device. The question arises: Could not software be constructed in the same fashion? Why are there not catalogs of software modules as there are catalogs of mechanical parts or VLSI devices? When developing a new piece of software, components should be selected from these catalogs and combined instead of reinventing the wheel over and over again. A technical difficulty of reusability is the fact that the problems in software design occur in small variations, they are *similar* but *not exactly* the same. This requires a proper amount of flexibility of the modules.

To some extent reusability has been applied successfully for software development in scientific computation. Excellent routine libraries like e. g. those from *NAG* and *IMSL* are commonly used for solving problems in linear algebra, differential equations, optimization and other fields. In these applications each problem admits a simple specification in terms of a small set of parameters and there are no complex datastructures involved. A similar reuse should occur on a higher level of abstraction where not only a certain functionality is involved but a *set* of functionalities, combined with sophisticated datastructures as it is typical for computer algebra software. As it will turn out, object oriented design is aimed at solving many of these problems.

2.2 Object-Oriented Design

Object. *A person or thing to which action, thought or feeling is directed. Anything visible or tangible; a material product or substance.*
Class. *A number of people or things grouped together because of certain likeness or common traits.*

Webster's New Twentieth-Century Dictionary, 1977

Object oriented design is not simply a new feature added to existing programming languages. Rather it is a *new way of thinking* about the process of decomposing problems and developing software tools for its solution [19]. It is founded on the fundamental idea that it is better to structure software by means of the objects than by the actions because it faciliates the enforcement of the principles described in the preceding section.

Objects and classes. There are two fundamental concepts in object oriented design: The *object* and the *class*. An object is an abstraction of something in the domain of a problem, reflecting the capabilities of a system to keep information about it or to interact with it. It is an encapsulation of *state* (data values) and *behavior* (attributes and operations). A class is a description of one or more objects, describable with a uniform set of *features*, i. e. its attributes and operations. The *interface* of a class provides its outside view, primarily it consists of the declarations of all features applicable to instances of this class. The *implementation* of a class is its inside view.

Initializing an operation with an object is expressed as *sending a message* to this object. The object reacts to the message by *executing a method.* The method describes which actions are actually performed on a given object. Objects that share the same methods belong to the same class. The association of messages and methods is determined by the respective class definitions and occurs at run time. This scheme differs basically from function or procedure invocations in conventional languages.

Object-oriented design OOD is a method of organizing software into a collection of objects. Each object is an *instance* of some class, the class of an object is its *type.* A class is a *static* concept, it is a unique piece of program text. In contrast to that, an object is a *dynamic* concept which occupies some space in computer memory at run time. The distinction is very much like that between a type and values of that type. The identification of classes and objects is the hardest part of *OOD.* The general guideline is the theory of data abstraction: A notion should only be made into a class if it describes a set of objects characterized by interesting operations and meaningful attributes, expressed in the terminology of the problem domain. In almost any nontrivial case, the design of classes and objects is an incremental, iterative process and it will hardly ever be possible to define them exactly right the first time. Creating too many, mostly unneeded classes is a typical novice's mistake. The essence of *OOD* may be summarized as

$$OOD = \text{Objects} + \text{Classes} + \text{Inheritance}$$

where the last keyword denotes a special kind of relationship among modules that is the subject of the next paragraph.

Relationships among classes. Classes do not exist in isolation. Rather, for a particular problem domain, the key abstractions are usually related in a variety of ways, forming the basis for the class structure of the software design. Most object-oriented languages support the following relations between classes.

- Inheritance relationships
- Using relationships
- Instantiation relationships

Inheritance defines a relationship among classes, wherein one class shares the structure or behavior defined in one or more previously defined classes called *single inheritance* or *multiple inheritance* respectively. A class may be defined as an extension or a restriction of another. Therefore it is called a *is_a* relationship. If A and B are classes, the specification A *is_a* B or A *inherits* B may be considered as an abbreviation mechanism that avoids redefining the features of class B in the definition of class A.

A class from which another class inherits is called *superclass* or an *ancestor.* Similarly a class that inherits from one or more classes is called a *subclass* or a *descendant.* Inheritance is always transitive, so a class can inherit features from superclasses many levels away. Inheritance means that the behavior and data associated with a subclass are always an extension, i. e. a strictly larger set, of the features of the superclass. All classes are members of a hierarchy which is generated by these inheritance relationships. If only single inheritance is supported, the class hierarchy is shaped as a tree or, more properly, a forest. If multiple inheritance is allowed, it is a directed acyclic graph (DAG).

In a complicated class hierarchy allowing multiple inheritance the following two problems may occur. *Name collisions* are possible when two or more different superclasses use the same name for an exported feature. The second problem is *repeated inheritance* which occurs if a class is an ancestor of another in more than one way. The question to be answered in these situations is this: What is the meaning of its features for the descendant? There are several solutions possible. The language semantics might regard it as illegal and reject the compilation of the class, it might require that all references to the name fully qualify the source of its declaration, or some sort of precedence mechanism might be applied in order to remove the duplicates.

Often inheritance is not adequate to capture the proper relationship among different classes. A class's interface or its implementation can refer to features defined in other classes. The former class is then called a *user* of the latter. A user imports features of another class, but internal details of this class remain protected againts access by the user. If a class B needs a facility from another class A, there arises the question as to what the proper choice of a relationship among these classes should be, and there may be no obvious answer. The following rule may be applied as a guiding principle: Inheritance is appropriate if every instance of B may be also viewed as an instance of A. The user relationship is appropriate if every instance of B simply needs a few features of A.

Polymorphism is intimately related to the notion of inheritance. One might even say that the power of object-oriented design is due to a large extent to the polymorphism it faciliates. This discussion follows closely [15]. In a polymorphic language some values and variables may have more than one type. A particular kind of polymorphism, usually called *ad-hoc polymorphism*, has been around for some time and is not closely bound with *OOD*. It comprises *overloading* and *type coercion*. In *overloading* the same variable name is used to denote different functions. The context has to be used in order to decide which function is denoted by a particular instance of the name. A *coercion* is a semantic operation that is needed in order to convert an argument to the type expected by a function. It allows the user to omit the necessary type conversion, it must be determined by the system and used by the compiler to generate the required type conversion code. According to Cardelli and Wegner ([15], page 477), ad-hoc polymorphism is some kind of *apparent* polymorphism whose polymorphic character disappears at close range.

On the other hand, *universal polymorphism* is considered true polymorphism. It materializes either in *inclusion polymorphism* or *bounded polymorphism* where every object of a subtype can be used in a supertype context, or in *parametric polymorphism* which means that the same object or function can be used uniformly in different type contexts without changes.

Another important mechanism that enhances the flexibility of an *OOD* are *abstract classes*. In an abstract class the operations are only declared, they are also called *deferred* or *abstract operation*. Its implementation is defined in some descendant which inherits this operation. One consequence is that an abstract class cannot be instantiated. This concept has an interesting application in the system *Axiom*.

Concept	Ada	Smalltalk	C++	CLOS	Eiffel	Axiom
Single inheritance	no	yes	yes	yes	yes	yes
Multiple inheritance	no	no	yes	yes	yes	yes
Generic classes	yes	no	no	no	yes	yes
Abstract classes	no	yes	no	yes	yes	yes
Dynamic objects	no	no	no	no	yes	yes
Concurrency	yes	yes	yes	yes	yes	no

Table 2: Computer languages that support various concepts from object-oriented design as of 1993. *Axiom* is especially designed for computer algebra applications.

A class that allows parameters in its definition is a *generic class*. The parameters are usually types. Genericity allows it to construct new classes by substitution of types. A generic class must be *instantiated* , i. e. its parameteres must be filled in before objects of this class may be created. Genericity is only meaningful in a typed language. An important application for genericity are *container classes*. Its objects are homogeneous collections of the same type. The most common kinds of container classes include sets, bags, queues and graphs.

The runtime environment of object-oriented systems. In order to be able to take advantage of the facilities provided by an object-oriented software system in its entirety, the runtime environment must provide the necessary support for it. The most important concepts are the *dynamic instantiation* of objects and the *dynamic binding*. The latter notion means that the run-time system automatically selects a method adapted to the corresponding object.

Dynamic object creation is one of the most useful features of object-oriented languages. As opposed to the static model where any piece of memory used by a program is allocated at loading time, in a fully dynamic model objects are created dynamically by explicit request at run time. This scheme is not specific to object-oriented languages, e. g. it exists also in *LISP* where the function CONS causes the dynamic creation of a pair (see the next section on *REDUCE*). However it is of particular importance in object-oriented languages and in computer algebra applications. When objects may be created at runtime, the problem arises of what to do when they are not used any more. In order to handle this problem some kind of *memory management* is needed. The easiest solution for the language implementer is to provide a *reclaim* function that tells the system that an object is no longer needed and its memory space may be recycled. A much better approach is to implement some kind of *garbage collection* that handles these occurences automatically and is available in most *LISP* based systems.

Comparison between object-oriented languages. The concept of object-oriented programming has its roots in *SIMULA 67* [16]. It was more fully developed in the evolution of the language *Smalltalk-80* [17] from which also a large portion of the terminology introduced in this section originates. In the meantime there are several dozens of languages that support *OOP* in some way or another. A listing of the more prominent representatives is given in Table 2. Of particular interest here is the language *Axiom* which has especially been designed for computer algebra applications.

3 A Conventional System: REDUCE

Although most of the design principles described in the last section have not been applied for developing *REDUCE*, it is helpful to think of its architecture in terms of datatypes and operations. This leads to the scheme shown in Figure 2. The uppermost level is the algebraic mode of *REDUCE*. Usually a novice starts working with the system at this level. It is described in the manual that is provided by the developer [2], see also the book by MacCallum [3] or the interactive tutorial [4].

In the course of time a serious user will realize that the possibilities of the algebraic mode are not sufficient for implementing sophisticated software by himself. In particular the access to the various datastructures, the facilities to manipulate them and to create new ones are too limited. In order to proceed it is absolutely necessary to get a better understanding of the internal workings of the system. According to the scheme of Figure 2 this means to have a closer look at the *Symbolic Mode* and also the *RLISP Mode*. They will be described subsequently. On top of these levels the user may develop the software according to his specifications with the functionality he needs in his applications. If this is performed in a systematic way, the first step of software development is to design and to implement an algebraic type system. Its position is indicated in bold face letters in Figure 2. In this scheme, the algebraic mode of *REDUCE* is mainly applied for input and some output purposes.

Abstraction Level	Data & Operations
Algebraic Mode	*Data:* Variables, Polynomials, Vectors, Matrices etc. *Operations:* Algebraic and Analytic Manipulations, Input & Output
	User Defined Algebraic Type System
Symbolic Mode	*Data:* Numbers, Kernels, Standard Forms & Quotients, etc. *Operations:* Simplification, Polynomial Arithmetic, Factorization, etc.
RLISP & SLISP	*Data:* Atoms & Pairs *Operations:* CAR, CDR, CONS, ATOM
Assembler	*Data:* Machine addresses *Operations:* Move, Load etc.
Object Code	*Data and Operations:* Bitstrings

Figure 2: The various layers of *REDUCE* organized by data types and operations.

3.1 The RLISP Mode of REDUCE.

LISP programs are usually characterized by a large amount of brackets over which control is easily lost. Therefore A. C. Hearn developed in 1973 a parser which allows it to write LISP programs in an ALGOL or PASCAL like syntax. The parser reads the RLISP expressions, translates them into LISP and evaluates them if there are no syntax errors. So RLISP provides the full possibilities of LISP in a syntax which might be more familiar to the majority of users. Certainly the 'R' which precedes the name LISP is supposed the remaind the user of its original purpose, i.e. to serve as convenient language for the computer algebra system *REDUCE*. The material described in this section has not as its aim to give a complete account of the RLISP system. Rather it should provide the necessary background for the subequent section on the symbolic mode.

Pairs and lists. The most general data type occuring in any LISP system is the *symbolic expression* or *S-expression* for short. Among others it comprises the four basic data types *identifier*, *integer*, *boolean* and *dotted pair* or simply *pair*. Identifiers and integers combined are called *atoms*. The shorthand *any* is used frequently to abbreviate the datatype description of an arbitrary S-expression. The two special identifiers T and NIL play an outstanding role whenever talking about programming in LISP. By definition the latter is absolutely equivalent to an empty pair of brackets. Therefore the atom NIL may always be replaced by () and vice versa, i. e. NIL is equivalent to () and (A . NIL) is equivalent to (A . ()).

The existence of pairs and lists is a distinguishing feature of any LISP system and plays a fundamental role in any application. They are therefore considered in more detail in this subsection. Dotted pair structures may be difficult to understand because its excess of parentheses and dots. Therefore list notation was invented to simplify reading and writing of complicated dotted-pair structures. The transformation rules which govern the transition from dotted pair to list notation may be expressed as follows.

1. Whenever a dot is immediately followed by a NIL, both may be omitted.
2. Whenever a dot is immediately followed by a "(", erase the dot, the "("and the corresponding closing ") ". The expression between the removed pair of brackets is left unchanged.

Repeated application to a given dotted pair structure terminates after a finite number of steps with a unique result. So any non-atomic S-expression may be represented in two equivalent ways, the *dot notation* and the *list notation*. According to these rules one has (A . NIL) is equivalent to (A), ((NIL . NIL) . NIL) is equivalent to ((NIL)), (A . (B . (C . NIL))) is equivalent to (A B C) and (A . ((B . C) . NIL)) is equivalent to (A (B . C)). The reversal, i.e. the transition from list notation to dot notation is governed by the two rules

1. Replace each pair "() "with NIL.
2. Place a dot to the right of each list element and enclose the remaining part of the list with a pair of brackets. If there is no element left, use an empty pair.

Applying these rules the notation in the examples given above may be reversed. Important subclasses of all non-atomic S-expressions are formed by the *lists* and the *proper lists* or *P-lists*. They are defined by

list = () | (any . *list*)

P-list = () | (atom . *P-list*) | (*P-list* . *P-list*)

To see whether a given datastructure is a proper list one has to proceed as follows. Remove as many dots and brackets as possible according to the rules given above. If this process terminates with an expression which does not contain any dot at all, it is a proper list. If it does not contain a dot at the uppermost level, it is a list. Lists are a subclass of all possible S-expressions as well as proper lists are a subclass of all lists. If the outermost brackets of a list are removed, in general several disconnected pieces will remain which will be atoms or pairs. They are called the *top level elements* of that list. Their number is called the *length* of the list. If a top level element is again a list, the same procedure may be repeated. The deepest nesting of lists is called the *depth* of the original list. If none of the top level elements of a list is an atom it is called an *a-list* which is a shorthand for *association list*. By this definition ((A . B) (C . D)) and ((A B)(C D)) are both a-lists.

Graphical representation of pairs. Whereas list notation originated from the desire to have a more convenient way of writing data structures in LISP, dotted-pair notation is superior to list notation in another respect. It is very close to the internal structure in which the machine represents LISP data. Often this results in a much better understanding of how LISP functions operate. For that reason a graphical representation of dotted pairs is introduced as a third alternative. A dotted pair is represented as a rectangle divided into a left and a right box. A number or a greek letter in front of that rectangle followed by a colon is the physical address of that pair which uniquely identifies it. The left box obviously corresponds to the CAR part of the pair, the right box to the CDR part. The contents of either box is another address in computer memory. It may be the location of another dotted pair or an atom. In the former case the respective box and the addressed pair are connected by an arrow. In the latter the name of the corresponding atom is written into the box because atoms are always uniquely represented in computer memory. The important point to remember is that in any case a box contains information on a unique place in computer memory. If the content of a memory cell is the address of some other location in computer memory it is called a *pointer* to that place. Some examples for the graphical representation of LISP data are

(A . B) is represented by

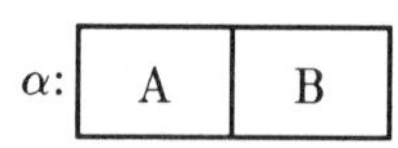

(A B) is represented by

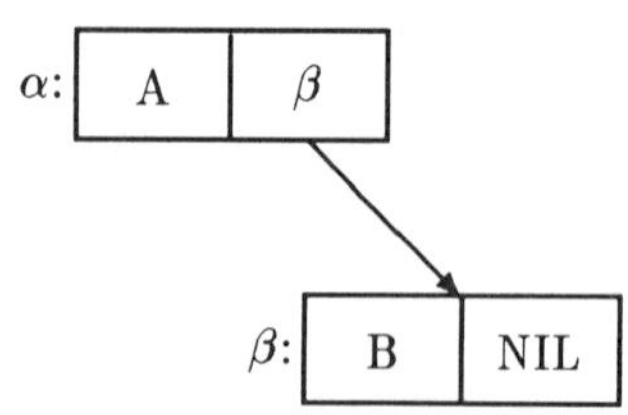

((A . B) . (A . B)) is represented by

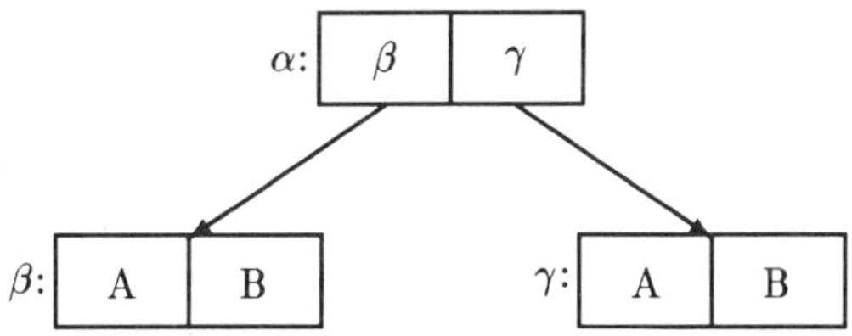

((A . B) . (A . B)) is represented by

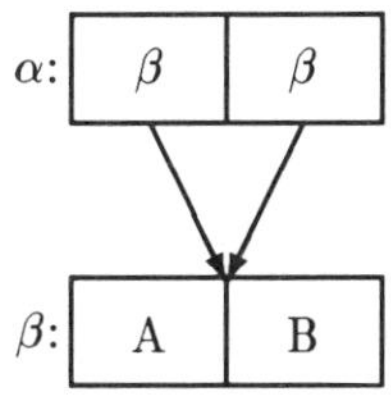

The first example represents simply a dotted pair whereas the second is a list. The latter two examples show that there are datastructures which look exactly alike when they are printed and still are rather different when their internal representation is considered, i.e. they are located at different places in computer memory. From this graphical representation another important property of LISP datastructures is clearly seen, i.e. that it permits *overlapping.* That means a single cell may be shared by several data structures.

Elementary RLISP functions: ATOM, CAR, CDR and CONS. The basic RLISP system function for making a distinction between atoms and pairs is the predicate ATOM. It is applied according to the syntax

ATOM(U:any): boolean

Its single argument is evaluated once. If the result is an atom, the answer T is returned and NIL otherwise. A few simple examples for using them are

ATOM'X; ⟶ T, ATOM 3; ⟶ T, ATOM'((A 1) (B 2)); ⟶ NIL

The basic functions for breaking up data structures into smaller pieces are the functions CAR and CDR with the syntax

CAR(U:pair):any and CDR(U:pair):any

When applied to a pair, the effect of CAR and CDR is to break it up and to return its left part or its right part respectively as it is shown by

CAR'(A . B) ⟶ A, CDR'(A . B) ⟶ B, CAR'(A B) ⟶ A, CDR'(A B) ⟶ (B)

The difference between the action on dotted pair (A . B) and the list (A B) is obvious. Because of its frequent occurence, it is advantageous to describe the action of CAR and CDR on lists without relying on the dotted pair representation. It is governed by the following rules. To obtain the CAR value of a list, remove the outermost pair of brackets and omit everything except the leftmost isolated element. To obtain its CDR value, again remove the outermost pair of brackets, remove the leftmost isolated element and attach the brackets around the rest. A few examples for applying these rules are

CAR'(A B C); ⟶ A, CDR'(A B C); ⟶ (B C), CADR'(A B C); ⟶ B

The basic operation generating dynamically new datastructures is the function CONS with the syntax

CONS(U:any,V:any):pair

At first both arguments are evaluated once from left to right. Then a free LISP cell is chosen, the address of the value of U is written into the left part and the address of the value of V into the right part. The address of this free cell is the root of the binary tree structure for CONS(U,V) which may be written now as (U . V). A basic feature of CONS is that it does not change existing structures but it creates new ones, so the two arguments remain unchanged upon calling it. The action of CONS may be seen from the following calling sequence

U:='A; ⟶ A, V:='B; ⟶ B, W:=CONS(U,V); ⟶ (A . B), W; ⟶ (A . B)

In more complicated applications it is usual to apply the infix version of CONS which is simply a dot surrounded by two spaces as it is shown in the following two examples

'A . 'B . 'C; ⟶ (A B . C), ('A . 'B) . 'C; ⟶ (A . B) . C)

The two data structures which are created by these calls are completely different as it may be seen from the graphical representations

(A B . C) is represented by

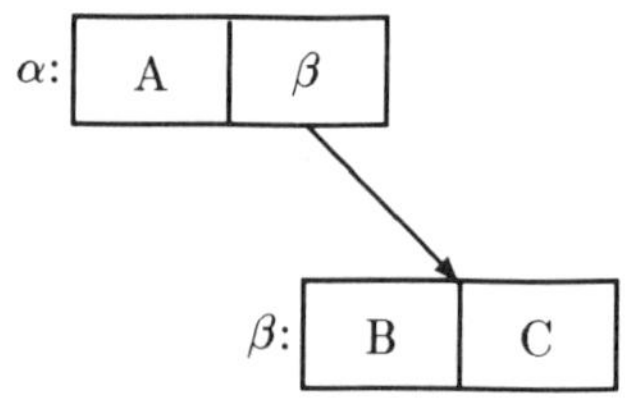

and

(((A . B) . C) is represented by

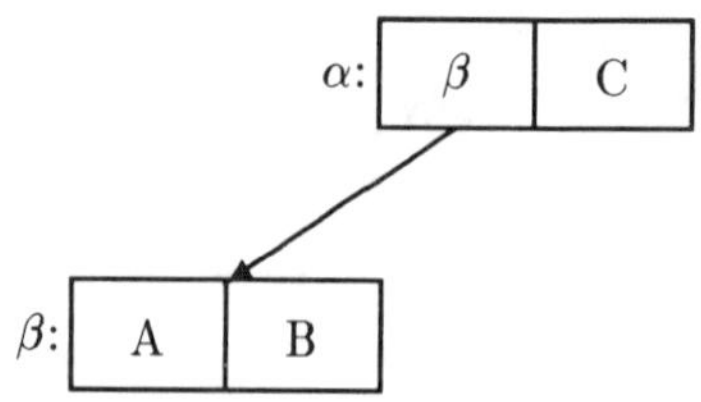

respectively. The concepts and the terminology which has been introduced in this subsection are an irrevocable prerequisite for understanding the next higher level in the hierarchy, i. e. the symbolic mode of *REDUCE*.

3.2 The Symbolic Mode of REDUCE.

The symbolic mode of *REDUCE* is characterized by the occurence of the new datatypes *standard form*, *distributive form* and *standard quotient*, provided with the operations like e. g. polynomial and rational function arithmetic which make them into mathematical objects. The building blocks out of which these objects are made are elements from a coefficent domain and various kinds of variables. They are implemented in terms of the datastructures and the operations defined in the RLISP system described in the preceding section.

Variables and coefficient domains. The coefficient domains which are supported by the *REDUCE* system are the ring $\mathbf{Z}$ of integers, the rational number field $\mathbf{Q}$, the ring of integers modulo n, $\mathbf{Z}_n$, or the real numbers $\mathbf{R}$ with a varying precision. These various domains are controled by the global switches *rational*, *modular* accompanied by the function *setmod* for defining the modulus n, and *rounded* accompanied by the function *precision* the integer argument of which specifies the number of valid digits. They are activated by means of the *on/off* declaration. Examples for its use are

on rational; *on modular; setmod 3;* *on rounded; precision 5;*

The other elementary constitutents for any polynomial are its variables. In *REDUCE* they are represented as *kernels.* With respect to the operations defined for polynomials or rational functions, they are considered to have no internal structure. In mathematical terms the variables may be just scalars like x, indexed quantities like x_1 or $y_{1,2}$ representing the components of a vector or a tensor respectively, or more structured objects like derivatives $df(x)/dx$ and $d^2f(x)/dx^2$. The general systematics may be seen from the formal definition in terms of the Backus normal form using the LISP dotted pair notation

<kernel>::=<identifier>|(<identifier> . <list of arguments>)

Here nonterminals are enclosed between a pair of brackets, productions are separated by a ::=-sign and alternatives by a |. From this definition it is obvious that a *REDUCE* kernel is either a LISP identifier or a list the first element of which is an identifier.

If a variable is a function of some other variable, i. e. if it is *dependent* on it, it is called a *derivative*, possibly of order zero. This situation is handled in *REDUCE* by the global variable *depl** of the type LISP *list list kernel.* Each entry of the list *depl** causes the first kernel to depend on the remaining ones. This action may be seen from the following sequence of input commands.

algebraic; *depend f,x,y;* *depend g,y;* *lisp;* *depl*;* $\longrightarrow$ *((f x y) (g y))*

From now on the kernels f and g are considered as being dependent on x and y or just y respectively. Its main effect is that the derivatives with respect to these variables do not vanish.

Standard forms and distributive forms. Multivariate polynomials may be regarded as polynomials in one main variable with coefficients which are polynomials in the remaining variables. This representation has the advantage that recursive algorithms may naturally be used for various operations. In *REDUCE* polynomials in the recursive representation are called *standard forms* which again are build up from *standard terms* and *standard powers.*

The most elementary building blocks are *domain elements* and *kernels.* Their interrelation is most easily described by defining these various quantities in Backus normal as above.

<standard form>::=NIL|<nonzero integer>| (<standard term> . <standard form>)

<standard term>::= (<standard power> . <standard form>)

<standard power>::=(<kernel> . <natural number>)

Considered as a LISP data object, a standard form in general is a pair with a special structure. There are special standard forms which are lists. A standard term and a standard form differ by its bracketing structure even if they represent identical mathematical objects. This is clearly seen from the *REDUCE* representation of the single term polynomial $2x^3$. The standard form and the standard term representation of this object are

(((X . 3) . 2))) and ((X . 3) . 2)

respectively. It is important to distinguish carefully between these two objects because all REDUCE system functions depend on that structure.

The internal representation of standard forms in REDUCE is best understood from the graphical representation in terms of LISP cells which has been introduced in the preceding subsection on RLISP. This representation is valid for a standard form which is neither NIL nor a domain element.

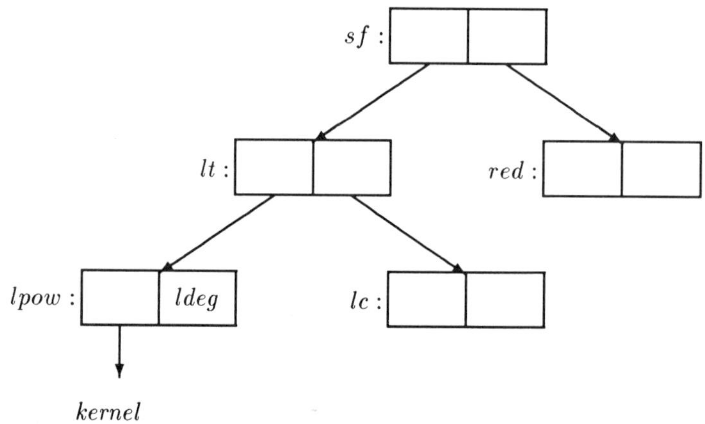

The abbreviations sf, lt, red, $lpow$, lc and $ldeg$ have the meaning standard form, leading term, reductum, leading power, leading coefficient and leading degree respectively. The special term *kernel* becomes clear from this representation. It is the most deeply nested object occuring in a polynomial which is considered as undivisible as far as operations on polynomials are concerned.

As a specific example of how a polynomial is represented internally consider the following univariate polynomial in x which has been assigned as value to the variable F in the algebraic mode.

F:=3∗X^2-2∗X+5;

Its standard form representation is obtained by calling

u:=lisp numr simp reval'f; $\longrightarrow$ (((X . 2) . 3) ((X . 1) . -2) . 5)

The three standard terms from which the polynomial F is built up are easily identified. Its graphical representation is obtained from the picture for the general case by appropriate substitutions and has the form

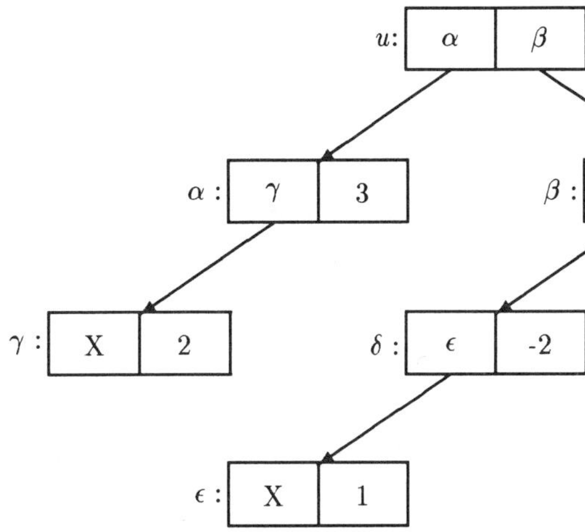

The most elementary operations on standard forms out of which all other operations are composed consist in selecting certain parts of it like the leading power, the leading term etc. Therefore they are called *selectors.* As a rule these operations require only a few LISP operations and are therefore performed in constant time. A large number of selectors is provided in the symbolic mode of *REDUCE.* The most important of them are listed in Table 3. Their names usually have some mnemonic meaning which is obvious in most cases and is considerably easier to remember than the combination of CAR's and CDR's which build them up.

Function Name & and Signature	Specification of Result
mvar(u:sf):kernel	main variable of *u*
lt(u:sf):lt	leading term of *u*
lc(u:sf):sf	leading coefficient of *u*
lnc(u:sf):de	leading numerical coefficient of *u*
ldeg(u:sf):integer	highest power of main variable
lpow(u:sf):	*mvar u* raised to *lpow u*
red(u:sf):sf	reductum
tvar(u:st):kernel	main variable in term *u*
tc(u:st):sf	coefficient of term *u*
tdeg(u:st):integer	degree of *tvar u*

Table 3: Some *REDUCE* selector functions for standard forms and standard terms.

Out of these selectors the functions defining the polynomial arithmetic are composed. As a rule, the cost of calling any of them is proportional to some low power of the input size which may be measured by the degree or the number of terms. The most important arithmetic operations for *REDUCE* standard forms are given in Table 4.

Function Name & Signature	Specification of Result
addf(u:sf,v:sf):sf	Sum of u and v
negf(u:sf):sf	Negative value of u
absf(u:sf):sf	u such that *lnc* u is positive
multf(u:sf,v:sf):sf	Product of u and v
quotf(u:sf,v:sf):sf	Exact quotient of u and v or *nil*
remf(u:sf,v:sf):sf	Remainder of u w.r.t. v
qremf(u:sf,v:sf):pair	Pair *quotf(u,v) . remf(u,v)*
expf(u:sf,n:integer):sf	u raised to power n

Table 4: *REDUCE* system functions defining the arithmetic of standard forms.

The next layer of operations are those for calculating greatest common divisors of standard forms and for factorizing them. They are shown in Table 5. As opposed to the functions considered so far, the cost for applying them may be substantial.

Function Name & Signature	Specification of Result
ckrn(u:sf):sf	Monomial gcd of u
gcdf(u:sf,v:sf):sf	Gcd of u and v
lcm(u:sf,v:sf):sf	Lcm of u and v
prim-part(u:sf):sf	Primitive part of u w.r.t. leading variable
fctrf(u:sf):list	complete factorization over the integers
sqfrf:(u:sf):list	square free factorization

Table 5: Operations for calculating gcd's of standard forms and for factorization

Although most operations on polynomials in *REDUCE* are performed in the standard form representation, there are exceptions from this rule where a distributive representation is more appropriate. This applies in particular to all Gröbner base calculations. In the distributive representation a polynomial is considered as being built up from monomials which are multiplied by elements from some coefficient domain. The structure of a completely distributive form is obvious from the following Backus normal form description of the corresponding *REDUCE* data structure

<distributive form>::= NIL|(<distributive term> . <distributive form>)

<distributive term>::=(<list integer> . <sq>)

The transformation between standard forms and distributive forms is achieved by the two functions

$f2df(u:sf):df$ and $df2sf(u:df):sf$

with an obvious semantics. They may be considered as type transformation functions between sf's and df's. In order to operate properly, the variables with respect to which a distributive representation is desired have to be assigned as a list to the global variable *zlist*. The arithmetic operations for distribtutive forms are given in Table 6.

Function Name & Signature	Specification of Result
plusdf(u,v:df):df	Sum of u and v
negdfdf(u:df):df	Negative value of u
multdf(u,v:df):df	Product of u and v

Table 6: Arithmetic operations for distributive forms.

The last objects to be dicussed in this section are rational functions which are represented in *REDUCE* as *standard quotients.* The datastructure that is applied for its internal representation is a LISP pair of two standard forms corresponding to its numerator and the denominator respectively. The general structure of a standard quotient is described by

<sq>::=<sf> . <sf>

The special value 0 corresponds to the pair (*nil* . 1). The elementary operations for composing and decomposing a standard quotient are given in Table 7. The last function *canonsq* cancels the gcd of the numerator and the denominator. Due to the possible cost of finding the gcd of the numerator and the denominator, the user is enabled to control this process explicitly. The operations defining the arithmetic of standard quotients are given in Table 8.

Function	Returned Result
numr(u:sq):sf	Numerator of u and v
denr(u:sq):sf	Denominator of u
(u:sf ./ v:sf):sq	Unsimplified quotient u/v
canonsq(u:sq):sq	Simplified form of u

Table 7: Functions for composing and decomposing a standard quotient.

Function	Returned Result
addsq(u:sq,v:sq):sq	Sum of u and v
negsq(u:sq):sq	Negative value of u
multsq(u:sq,v:sq):sq	Product of u and v
quotsq(u:sq,v:sq):sq	Quotient of u and v
invsq(u:sq):sq	Inverse of u
exptsq(u:sq,n:integer):sq	nth power of u

Table 8: *REDUCE* functions for defining the arithmetic of standard quotients.

This short description of some of the features of the symbolic mode of *REDUCE* shows that some useful structure may be obtained if one thinks of it in terms of the general concepts discussed in the last section. It should be contrasted with the descriciption of the system *Axiom* in the next section.

4 A 3rd Generation System: Axiom

The terminology which is applied in this section is a combination of mathematical notation *and* the object-oriented terminology from section 2. This is not accidental. After several attempts to utilize advanced know-how in software design for a computer algebra system (see e.g. Fedoraro's *Newspeak* [20]), *Axiom* of the computer algebra group of R. D. Jenks of IBM is the first general purpose computer algebra system that applies systematically the principles of object oriented design. It is therefore called a 3rd generation system. The other systems mentioned in this article belong to the 2nd generation. As a result, it provides an environment for handling mathematical objects in scientific computation with excellent structural features and almost unlimited expandability. It is designed to support an increasing functionality with a minimal increase of structural complexity. This section is *not* a substitute for the original literature on *Axiom* as e. g. the book by R. D. Jenks and R. S. Sutor [9] or the *Axiom User's Guide* [24]. Rather it is a short introduction, pointing out some essential features in the context of *OOP* and providing the necessary background for the subsequent sections.

4.1 Categories, Domains and Packages.

These three units are the basic building blocks out of which any piece of *Axiom* software is constructed. They form the constituents both of the *Axiom* system and the user developed software and form also the basic compilation units. The constructors of these units are described in a semi-formal way. This is *not* a description of the syntax, rather it is a means of describing a form that will be accepted by the compiler without claiming that this is the only possibility. For example, the term *terminal* does not necessarily imply that the respective string is a keyword of the language. In some cases it is just a suggestive placeholder and any other string would serve the same purpose. To this end the following conventions will be applied. Terminal symbols in the sense as it just has been explained are written in roman letters. In this connection the ==-sign may be considered as a terminal symbol. *Italics* denote non-terminals, they indicate the position of the program code into which the respective symbol has to be expanded according to the specification of the given constructor. The meta-symbols [,] and | are not part of the *Axiom* language, they serve to structure the description of the syntax. The text enclosed between a pair of brackets [and] is optional, alternatives are separated by a |. Text within braces { and } may be repeated any number of times including zero. The =-sign separates the left- and right- hand side of a production.

Categories are Second Order Types. Categories provide a set of generic operations together with a set of attributes describing the algebraic properties of these operations. In the terminology of *OOP*, a category is most closely approximated by an abstract class. By definition, the type of any category is Category. Categories are used to define interfaces of domains and packages of polymorphic functions. The syntax of a category constructor is partly defined by the following set of productions.

$$\begin{array}{rl} CategoryConstructor = & CategoryName\text{:Category} == FeatureList \\ & [\text{add } [ImportList]\ Defaults] \end{array}$$

```
-----------------------------------------------------------------------------
|Group(): Category == Monoid with                                             |
|    inv:  $     -> $                 ++ inv(x) returns the inverse of x      |
|    "/":   ($,$) -> $                ++ x/y is the same as x*inv(y)          |
|    "**": ($,Integer) -> $           ++ x**n returns x raised to the power n |
|    conjugate: ($,$) -> $            ++ conjugate(p,q) computes inv(q)*p*q   |
|    commutator: ($,$) -> $           ++ commutator(p.q) computes             |
|                                     ++ inv(p)*inv(q)*p*q                    |
|  add                                                                        |
|    import RepeatedSquaring($)                                               |
|    x:$ / y:$ == inv(x)                                                      |
|    conjugate(p,q) == inv(p) * p * q                                         |
|    commutator(p,q) == inv(p) * inv(q) * p * q                               |
-----------------------------------------------------------------------------
```

Figure 3: The implementation of the system category Group.

$CategoryName = \;\; Name[(CategoryTypelist)]$

$FeatureList = \;\; Inherits \mid \text{with } Extensions \mid Inherits \text{ with } Extensions$

$Inherits = \;\; CategoryName \mid$

$\qquad \text{Join}(CategoryName, CategoryName\{,CategoryName\})$

$Extensions = \;\; \{Signature\}, \{Attribute\}$

$ImportList = \;\; \text{import } DomainName|PackageName\{DomainName|PackageName\}$

$Defaults = \;\; \{FunctionDefinition\}$

From these rules it is obvious that a category is either a collection of inherited features, of features which are newly defined in this category or both.

The rest of the syntax definition and the semantics of *CategoryConstructor* will be given informally. A *CategoryName* is an identifier which which by convention begins with a capital letter. If it specifies a category without parameters, this is already the complete description. Otherwise they are given in *CategoryTypelist*, enclosed within parentheses and separated by commas. The parameters of a category must be domains.

The *Inherits* part of the *FeatureList* specifies from which already existing categories the currently defined one inherits. If there is only a single ancestor, only its name has to be given. If there are two or more, they have to appear as arguments of the function Join. It returns the union of the operations and attributes of its arguments. In *Extensions* the features which are newly defined are specified by listing the signatures of the operations and the attributes in the form

$Signature = FunctionName\text{: } TypeList \; - > Type$

$TypeList = ()|Type|(Type\{, Type\})$

$Attribute = Name[(Signature)]$

Each *FunctionName* is the name of an operation, or an infix operator enclosed in quotation marks. A *Type* is a domain or a \$-sign. The latter will be explained in connection with the usage of categories as interfaces of domains in the subsequent paragraph.

In the optional *Defaults* branch those operations are defined which are true in any domain belonging to that category. Those domains or packages which are used for this implementation are included in the *ImportList*.

An example of a category the implementation of the system category Group is shown in Figure 3. It does not have a parameter. The $-sign has always the meaning Group in this example. The category Group inherits from Monoid and uses some features from the package RepeatedSquaring. The text following the ++ in any line is a comment.

Domains are Types. Domains provide specific functions which implement the operations and satisfy the attributes of a certain category. In the sense of *OOP*, a domain corresponds to a class, a domain constructor corresponds to a generic class. It describes the computational objects which are obtained by its instantiation. The type of an object is defined by the domain from which it has been instantiated. The type of a domain is the category to which it belongs, its category provides an abstraction of its properties independent of its actual implementation. This is the reason why categories are called *second order types*. Domains are created by functions called *domain constructors*. Its syntax is partially described by the following productions.

DomainConstructor =	*DomainName*:Exports == Implementation where
	[*TypeDeclarations*]
	Exports == *FeatureList*
	Implementation == [*AddDomains*] add
	[Rep:=*Representation*]
	[*ImportList*]
	[*FunctionDefinitions*]

DomainName = *Identifier*[(*DomainTypelist*)]

The production rule for the new domain constructor with the name *DomainName* shows that it may take zero or more parameters. The parameters of a domain are either domains or objects. For reasons of clarity, the types of the formal parameters are often omitted or abbreviated in the domain name. In this case this information has to be provided by the *TypeDeclarations* which is defined by

TypeDeclarations = { *Type* | *Identifier* ==> *Identifier* }

The Exports branch defines the user interface of the newly defined domain in terms of the *FeatureList*, i. e. it defines the set of features to be exported by the domain. This is also expressed by saying that the newly constructed domain is an unconditional member of all categories which are listed in the *Inherits* branch of *FeatureList*. It has the same syntax and semantics as in the category constructor. A $-sign which occurs in the signatures of the *Extensions* branch of *FeatureList* indicates the type of the domain being currently defined. There must be at least one $-sign in a signature in order for the domain to define a new type. Programs which create and manipulate objects of a domain must do so only through the use

```
--------------------------------------------------------------------------
|SquareMatrix(n,R): Exports == Implementation where                       |
|    n: PositiveInteger                                                   |
|    R: Ring                                                              |
|  Exports == Join(Algebra(R),Ring,BasicMatrixCategory) with              |
|    trace: $ -> R          ++ trace(x) computes the trace of x           |
|    diagProd: $ -> R       ++ diagProd(x) computes the product of the    |
|                           ++ diagonal elements of x                     |
|  Implementation == RectangularMatrix(n,n,R)                             |
|    add                                                                  |
|    Rep:= Vector(Vector(R))                                              |
|    trace(x) == +/[x.i.i for i in 1..n]                                  |
|    diagProd(x) == */[x.i.i. for i in 1..n]                              |
--------------------------------------------------------------------------
```

Figure 4: The implementation of the domain SquareMatrix.

of operations exported by the domain. This description of a domain constructor makes it clear that a domain is the implementation of an abstract data type as explained above.

The purpose of the Implementation branch is to assure that all exported features of the domain are actually available. This may be achieved either by explicitly including it into the *FunctionDefinitions*, by inheriting it from a previously defined domain which is given in the option *AddDomain*, or by taking a default implementation from one of the categories to which it belongs. If a feature occurs in several of these places, first preference is always given to the local definition, an implementation which is also inherited is overwritten by the current definition. If there is no implementation found in any ancestor domain, a default in one of the categories to which the current domain belongs must be given and is adopted from there. Otherwise an error occurs. Accordingly the implementation of a feature may take place many levels away from the currently defined domain.

The internal representation in terms of lower level data types is specified by *Representation*. It is assigned to the keyword Rep. If no assignment to Rep is made, the internal representation is inherited from a previously defined domain in the *AddDomains* branch. In addition to implementing exported features, the *FunctionDefinitions* part provides also the operations that are locally needed. They are not visible to the outside because they are not included into the category part. The same is true for the internal representation. In order to make features from other parts of the system locally available within a domain constructor, the constructor from which it originates may be included in the *ImportList* which follows the Rep so that its features may be referred to generically. In this way a using relationship is established similar as it occured in the *Defaults* part of a category constructor.

An example of a domain part of the implementation of SquareMatrix is shown in Figure 4. It has two parameters. The first is an object with type integer, the second is a domain of type Ring. Its interface comprises the union of the categories Algebra(R), Ring and BasicMatrix-Category, supplemented by the two operations *trace* and *diagProd*. The implementations are either taken from the already existing domain RectangularMatrix in the *AddDomain* branch or they are explicitly coded in the Implementation part. The internal representation is a vector of vectors.

```
------------------------------------------------------------------------------
|Ratint(F,v): Exports == Implementation where                                |
|    F: Field | characteristc$F = 0                                          |
|    v: Expression                                                           |
|    UP: UnivariatePoly(v,F)                                                 |
|    RF: QuotientField UP                                                    |
|    LogList == List Logterm where                                           |
|                 Logterm == Record(coeff:NewF,logand:NewP)                   |
|                 NewF == UnivariatePoly("z",F)                               |
|  Exports == with                                                           |
|    integrate: RF -> Record(ratpart:RF, logpart: LogList)                   |
|  Implementation == add                                                     |
|    ++ here follows the implementation of the functions provided by        |
|    ++ the package                                                          |
------------------------------------------------------------------------------
```

Figure 5: Part of the implementation of the package Ratint for integration rational functions.

Packages are Collections of Generic Functions. A package may be used to extend the operations from a domain, or to put together several domains to achieve a new programming goal. Packages are special cases of domains inasmuch as a package does not define a new type. The difference is most obvious by comparing the domain constructor function with the package constructor that is given next.

PackageConstructor =	*PackageName*:Exports == Implementation where
	[*TypeDeclarations*]
	Exports == with *Extensions*
	Implementation == add
	[*FunctionDefinitions*]

PackageName = *Identifier*[(*PackageTypelist*)]

Here *TypeDeclarations*, *Extensions* and *FunctionDefinition* have the same syntax as defined previously. A semantic difference is that there does not occur a \$-sign in any signature because a package does not define a new datatype. For the same reason there is no internal representation and consequently no assignment to the variable Rep. A package constructor can take any number of types as parameters which are either objects or domains. The *FunctionDefinitions* part implements the exported and the local functions defined by this package.

As an example the package Ratint is shown in Figure 5. It provides the single operation *integrate* which integrates a rational function in a single variable over a field of characteristic zero. The answer requires a complicated data structure in order to store the answer in any case that may occur. It should be noted that there does not appear any \$-sign in this constructor.

4.2 Axiom System Constructors

In this section the distinction between a symbol-oriented and a math-oriented system will become particularly clear if the terminology is compared with that of section 3 where the *REDUCE* system has been described. There will be no mention made of anything like LISP objects as e. g. lists and pairs. Rather the description of the parametrized classes which are provided by the *Axiom* system will be given in a terminology that is mostly mathematical.

The first prerequisite for applying a system like *Axiom* in order to solve mathematical problems is a detailed description of the categories, domains and packages which are provided by it. In particular this means to give a survey of the system constructors, their interdependencies and their public parts in terms of which the communication with them is achieved.

Category Name	Abbreviation	Inherits from	New Features
AbelianGroup	ABELGRP	CancellationAbelianMonoid	* −
AbelianMonoid	ABELMON	AbelianSemiGroup	* Zero zero?
AbelianSemiGroup	ABELSG	SetCategory	* +
Finite	FINITE	SetCategory	index lookup random size
Group	GROUP	Monoid	** / commutator conjugate inv
Monoid	MONOID	SemiGroup	** One one? recip
OrderedAbelianMonoid	OAMON	AbelianMonoid OrderedAbelianSemiGroup	
OrderedAbelianSemigroup	OASGP	AbelianMonoid OrderedSet	
OrderedFinite	ORDFIN	Finite OrderedSet	
OrderedMonoid	ORDMON	Monoid OrderedSet	
OrderedSet	ORDSET	SetCategory	< max
Ring	RING	LeftModule Monoid Rng	
Rng	RNG	AbelianGroup SemiGroup	
SemiGroup	SGROUP	SetCategory	* **
SetCategory	SETCAT	CoercibleTo Object	=

Table 9: A Selection of *Axiom* System Categories in Alphabetical Order.

Axiom Categories. In Table 9 a selection of categories which are made available by the *Axiom* system is shown, alltogether there are almost 200. The name of each entry is accompanied by its abbreviation which is also known to the system. In the third column those categories are listed from which the current entry inherits one or more features. For a category inheritance means that a signature or an attribute is received from a previously defined category. Furthermore it may comprise the default definition of an operation. Inheritance relations generate a dependency structure a graphical representation of which may be found in the inside front cover of the book by Jenks and Sutor [9]. All inheritance relations may be obtained from the information given in Table 9. It may occur that a category does not define any new features, i. e. it is just a collection of features inherited from already existing categories as e. g. OrderedFinite. The reason is that such a category defines an interesting mathematical structure with the consequence that it may be applied in more than one place somewhere else in the system.

The only category that does not inherit any mathematical feature is the last entry SetCategory. It provides the only feature =. By inheritance it is available to all categories in the system. This may happen in several steps, e. g. the category Ring inherits the =-function over the four steps

SetCategory ⟶ SemiGroup ⟶ Monoid ⟶ Rng ⟶ Ring

A certain feature may be inherited from some other category in more than a single way as it may be seen from the following two paths

SetCategory ⟶ AbelianSemiGroup ⟶ AbelianMonoid ⟶ OrderedAbelianMonoid

SetCategory ⟶ OrderedSet ⟶ OrderedAbelianMonoid

```
----------------------------------------------------------------------------
|Group is a is a category  constructor                                     |
|Abbreviation for Group is GROUP                                           |
|------------------------------- Operations -------------------------------|
|?*? : ($,$) -> $                       ?**? : ($,Integer) -> $            |
|?/? : ($,$) -> $                       ?=? : ($,$) -> Boolean             |
|1 : () -> $                            coerce: $ OutputForm               |
|commutator : ($,$) -> $                conjugate : ($,$) -> $             |
|inv : $ -> $                           one? : $ -> Boolean                |
|recip : $ -> Union($,"failed")                                            |
----------------------------------------------------------------------------
```

Figure 6: The user interface of the category Group.

The system has to assure in those cases the uniqueness of this particular feature as it has been discussed in the section on object-oriented design. From Table 9 all features of any category listed there may be obtained by taking the transitive closure of all inheritance relations involved. As an example consider the collection of features of the category Group. A simple consideration leads to

=, *, **, /, One, one?, recip, commutator, inv

In general it is very time consuming and error prone to extract this kind of information out of a listing like the one given in Table 9. The *Axiom* system therefore provides tools which make it possible to obtain this information automatically. The answer which is obtained for the category Group is shown in Figure 6. It provides some additional information like the signatures of all functions. The $-sign in this table abbreviates the currently considered category, i. e. the Group constructor.

Axiom Domains and Packages. The mathematical objects that occur in the course of solving problems are instantiations of domains, they have the type of the domain to which they belong. Domains have therefore a more direct connection to the problems of the user than categories. More than 200 of them are defined and supported by the *Axiom* system. Packages

Domain Name	Abbreviation	Category Membership
DistributedMultivariatePolynomial	DMP	Polynomial
DifferentialSparseMultivariatePolynomial	DSMP	DifferentialPolynomialCategory
GeneralDistributedMultivariatePolynomial	GDMP	PolynomialCategory
MultivariatePolynomial	MPOLY	PolynomialCategory
OrderlyDifferentialPolynomial	ODPOL	DifferentialPolynomialCategory
Polynomial	POLY	PolynomialCategory
SparseMultivariatePolynomial	SMP	PolynomialCategory

Table 10: A Selection of *Axiom* System Domains Defining Various Polynomials.

are implementations of algorithms for already existing datatypes. The system provides about 300 of them covering a diversity of topics extending from algebraic functions, elementary function integration and various Gröbner base packages.

A selection of various polynomial domains is shown in Table 10. In the entry Category Membership only the most comprehensive category is given to which the respective domain belongs. It is considered as its (second order) type.

In order to use a system domain for software development, much more detailed information is needed about its interface. In particular it comprises the number and type of its arguments and the features which are made available by it. A typical example is the domain Sparse Multivariate Polynomial that is shown in Figure 7. Again a $-sign denotes an object of type SMP, a ?-sign indicates the position of an argument in an infix operator. The names of most operations have a mnemonic meaning so that in conjunction with the signature its specification is obvious. There is no mention whatsoever of the internal representation of objects of type SMP because it is not necessary to know it in order to use its functionality. This information is only available by looking at the implementation in the source file in complete agreement with the principles mentioned in section 2.1.

5 Developing Computer Algebra Software

In this final section it will be shown by means of various worked out examples how top quality computer algebra software may be designed in a *system independent* way. The basic prerequisites for it are a language and a type system. Both must be powerful enough to enable the implementation of sophisticated mathematical algorithms. At the same time they must be simple enough such that they are easy to learn, and it must be possible to understand previously implemented software easily.

5.1 General Principles

It is the purpose of this subsection to define an additional level of abstraction between the mathematical language and the language that a machine understands. To this end, a *computer algebra type system* and a *computer algebra language* are described. Based on these two concepts, mathematical problems may be formulated in a system independent way, i. e. its

```
------------------------------------------------------------------------------
|SMP(R: RING,VarSet: ORDSET)  is a domain constructor                         |
|Abbreviation for SparseMultivariatePolynomial is SMP                         |
|------------------------------- Operations -------------------------------|
|?+? : ($,$) -> $                        ?*? : (R,$) -> $                     |
|?*? : ($,$) -> $                        ?*? : ($,$) -> $                     |
|?*? : (I,$) -> $                        ?**? : ($,NNI) -> $                  |
|?-? : ($,$) -> $                        -? : $ -> $                          |
|?=? : ($,$) -> B                        associates? : ($,$) -> B             |
|coerce : R -> $                         coerce : I -> $                      |
|const : $ -> R                          constant? : $ -> B                   |
|content : ($,VarSet) -> $               degree : ($,VarSet) -> NNI           |
|degree : ($,L VarSet) -> L NNI          eval : ($,VarSet,R) -> $             |
|eval : ($,VarSet,$) -> $                eval : ($,L VarSet,L $) -> $         |
|lcm : ($,$) -> $                        leadingCoef : $ -> R                 |
|leadingMonomial : $ -> $                leadingPrimitiveMonomial : $ -> $    |
|likeUniv : ($,VarSet) -> SUP $          listCoef : $ -> L R                  |
|mindeg : ($,VarSet) -> NNI              mindeg : ($,L VarSet) -> L NNI       |
|monom : (VarSet,NNI,$) -> $             monom? : $ -> B                      |
|multipol : (SUP R,VarSet) -> $          numberOfMonomials : $ -> I           |
|pderiv : ($,VarSet) -> $                reductum : $ -> $                    |
|resultant : ($,$,VarSet) -> $           retract : $ -> R                     |
|retractable? : $ -> B                   totalDegree : $ -> NNI               |
|trailingCoef : $ -> R                   unipol : $ -> SUP R                  |
|unit? : $ -> B                          unlikeUniv : (SUP $,VarSet) -> $     |
|varlist : $ -> L VarSet                 varPol : VarSet -> $                 |
|LeadingCoef : $ -> R                    1 : () -> $                          |
|TrailingCoef : $ -> R                   0 : () -> $                          |
|content : $ -> R if R has GCDDOM                                             |
|?exquo? : ($,$) -> Union($,"failed") if R has INTDOM                         |
|gcd : ($,$) -> $ if R has GCDDOM and not R: RING = I and not R: RING = RN    |
|mainvar : $ -> Union(VarSet,"failed")  maxNorm : $ -> R if R has ORDRING     |
------------------------------------------------------------------------------
```

Figure 7: Part of the user interface of the *Axiom* domain Sparse Multivariate Polynomial is shown. This information may be obtained by the user interactively during an *Axiom* session.

description on this level of abstraction contains only the types and function names defined for them, and the constructs of this high-level language. In the current version the language supports only *programming in the small.* A complete version should also provide language constructs for *programming in the large*, i. e. creating new types, controling inheritance and usage of other modules etc.

A language for computer algebra. Programming languages are the mediator for communication between humans and computers, they transfer the representation of a mathematical problem into a form that can be understood by a computer. Programming languages characteristics can profoundly affect productivity of software manufacture and the quality of the result. A problem adapted special purpose language applies the terminology of the respective field of application and takes into account the conventional way of thinking and working in it. Its operations cover the basic algorithms for dealing with subproblems as large as possible in order to achieve a lot with very few statements. In other words, a specialized language must be well adapted to the structure of the problems to be handled in order to raise the efficiency of the user [21]. It is characterized by syntactic forms that have been especially designed for a particular application. They represent control and data structures at a higher level of abstraction than general purpose languages and eliminate thereby the need to specify algorithmic details. The know-how necessary for designing a high-level specialized language is highly domain-specific.

In computer algebra the data to be dealt with are mathematical objects. They are sets in the first place which may be provided with additional relations that make them into algebraic or ordered structures. Therefore it is not surprising that a language for computer algebra is modeled around set theory as it is true for the language *SETL* [23]. Moreover it is influenced by some properties of the *Axiom* language.

The fundamental syntactic units of our language are the *FunctionConstructor* and the *IterationConstructor*. They are partially described by the following set of productions where keywords are given in boldface letters.

FunctionConstructor = *FunctionName*(*Expression*|*Predicate*, *Iterator* {,*Iterator* })

FunctionName = **applies** | **select** | **exists** | **satisfies** | **separate**

Iterator = *Variable* **from** {**from**} *Set* | *Variable* **from** *Integer*,..,*Integer* [**by** *Integer*]

IterationConstructor = **iterate**(**true?** *Predicate*,*Expression*{,*Expression*}) |

iterate(*Expression*{,*Expression*},**true?** *Predicate*)

The remaining part of the syntax and the semantics will be explained informally. An *Expression* is any piece of code that is syntactically and semantically accepted by the interpreter or compiler and returns a result. In *Axiom* it is essentially an expression, in a LISP system it is very similar to a *form*. The free variables occuring in an *Expression* or a *Predicate* are evaluated in the environment in which the *FunctionConstructor* is called. All local variables are attached to certain sets by the subsequent *Iterator*.

The *Variable* in the first production alternative for the *Iterator* determines a variable name which may occur in the *Expression*, it specifies an element of the *Set*. Each occurence of the keyword **from** determines a level of nesting of *Set*. If there is a single occurence, the iteration occurs over elements of a set, if there are two, elements of a set of sets are taken, and so forth. If there is more than a single iterator, the iteration takes place over several sets, set of sets etc. in parallel. Both the number of iterators and the depth of nestings of sets are arbitrary up to the following constraints. The level of nestings in each iterator must be the same, the same is true for the cardinality of sets at each level. The second production for the *Iterator* allows it to iterate over intervals of integers, the stepsize being determined by the optional *Integer* following the keyword **by**.

The **applies** constructor causes the *Expression* to be evaluated for each selection of variables as determined by the *Iterator*. It returns an object with the same structure as it occurs in the latter with the result of the evaluations taken at the respective positions. The objects may be sets, sets of sets etc.

In the **select** and the **exists** constructors the *Expression* must be a predicate. The returned object of the **select** constructor has the same structure as those occuring in the iterators, it contains those elements of the *Set* occuring in the first *Iterator* for which the *Expression* evaluates to true. The **exists** constructor returns only the first element occuring in this *Set* for which the *Expression* is true. If the *Expression* begins with the keywords **maximal** or **minimal**, the remaining part of it must evaluate to an integer. In this case the **select** constructor returns all elements from the *Set* occuring in the first *Iterator* for which the *Expression* returns the maximal or minimal value respectively. The same applies to the **exists** constructor except that only the first element with this property is returned. A meaningful application will presume the proper number of occurences in either case.

The **satisfies** constructor is actually a predicate, it returns true if all evaluations of its first argument which must be a predicate also evaluate to true.

The **separate** constructor generates new sets of objects from the *Set* which occurs in the first *Iterator*. The newly created sets are formed such that the value of the *Expression* is the same for all elements of its member sets.

The two versions of the *IterationConstructor* are the basic syntactic unit for an unspecified number of repetitions of an action. In the first version, the *Predicate* is evaluated before any other action is taken. If the result is true, the *Expression*'s are evaluated from left to right one after the other. This proceeding is repeated until the *Predicate* evaluates to something that is not true. In this case the result of the last evaluation in the *Expression* is returned. If the *Predicate* is not true at the first evaluation, the value of the complete *IterationConstructor* is false. In the second version, a similar evaluation scheme is applied except that the *Predicate* is evaluated at the end of each round. The returned value of the constructor is always that of the last evaluation in the *Expression*.

In order to gain some familiarity with these language constructs, its usage will be illustrated by means of a few very simple, but typical examples subsequently. Assume the following assignments to the variables u, v, w and k have been made.

$$u := \{1, 0, -1, 3, 4, -4\}, \quad v := \{-1, 2, 3, -4, 2, 4\}, \quad w := \{u, v\}, \quad k := 1$$

The variable w actually denotes a set of sets of integers, it may also be considered as the representation of a matrix of integers. The following applications of the constructor **applies** are fairly obvious.

$$\textbf{applies}(abs\ \ x, x\ \textbf{from}\ u) \longrightarrow \{1, 0, 1, 3, 4, 4\}$$

$$\textbf{applies}(abs(x+k), x\ \textbf{from}\ u) \longrightarrow \{2, 1, 0, 4, 5, 3\}$$

$$\textbf{applies}(abs(x+y)-k, x\ \textbf{from}\ u, y\ \textbf{from}\ v) \longrightarrow \{-1, 1, 1, 0, 5, -1\}$$

The next example shows an application of an *Iterator* including a twofold nesting. It may be generalized to an arbitrary level of nestings by suitably generalizing the input.

$$\textbf{applies}(abs(x+k), x\ \textbf{from from}\ w) \longrightarrow \{\{2, 1, 0, 4, 5, 3\}, \{0, 3, 4, 3, 3, 5\}\}$$

The **select** constructor serves the purpose to extract certain sets by applying the predicate provided as the first argument.

$$\textbf{select}(x \neq 0, x \textbf{ from } u) \longrightarrow \{1,-1,3,4,-4\}$$

$$\textbf{select}(y > 0, x \textbf{ from } u, y \textbf{ from } v) \longrightarrow \{0,-1,4,-4\}$$

The next two examples show how different effects may be achieved by changing the sequence of the arguments in the *Iterator* of the **select** constructor.

$$\textbf{select}(x + y > 0, x \textbf{ from } u, y \textbf{ from } v) \longrightarrow \{0,-1,4\}$$

$$\textbf{select}(x + y > 0, x \textbf{ from } v, y \textbf{ from } u) \longrightarrow \{2,3,2\}$$

The **select** constructor works for any level of nestings of Set, here a twofold nesting is applied.

$$\textbf{select}(x > 0, x \textbf{ from from } w) \longrightarrow \{\{1,3,4\},\{2,3,3,4\}\}$$

In the following examples **select** is used with the special keywords **maximal** and **minimal**.

$$\textbf{select}(\textbf{maximal } abs\ x, x \textbf{ from } u) \longrightarrow \{4,-4\}$$

$$\textbf{select}(\textbf{minimal } abs(x+y), x \textbf{ from } u, y \textbf{ from } v) \longrightarrow \{1,-4\}$$

$$\textbf{select}(\textbf{maximal } abs(x-y-k), x \textbf{ from } v, y \textbf{ from } u) \longrightarrow \{-4\}$$

The **exists** constructor works similar except that it returns only single elements occuring within the set of its first *Iterator*. A few examples for its use are as follows.

$$\textbf{exists}(x > 0 \ \ and \ \ remainder(x,2) > 0, x \textbf{ from } u) \longrightarrow 3$$

$$\textbf{exists}(y > 0, x \textbf{ from } u, y \textbf{ from } v) \longrightarrow 0$$

The next two examples differ only by the selection of the order of the *Iterator*'s

$$\textbf{exists}(x + y > 0, x \textbf{ from } u, y \textbf{ from } v) \longrightarrow 0$$

$$\textbf{exists}(x + y > 0, x \textbf{ from } v, y \textbf{ from } u) \longrightarrow 2$$

In the following examples **exists** is used with the special keywords **maximal** and **minimal**.

$$\textbf{exists}(\textbf{maximal } abs\ x, x \textbf{ from } u) \longrightarrow 4$$

$$\textbf{exists}(\textbf{minimal } abs(x+y), x \textbf{ from } u, y \textbf{ from } v) \longrightarrow 1$$

The **satisfies** constructor is applied if certain properties are to be tested not only for individual elements but for complete sets as it is shown by the following examples.

$$\textbf{satisfies}(x \neq 0, x \textbf{ from } u) \longrightarrow false$$

$$\textbf{satisfies}(x \neq 0 \ or \ y \neq 0, x \textbf{ from } u, y \textbf{ from } v) \longrightarrow true$$

It works also for a nesting of sets higher than one as in the example

$$\textbf{satisfies}(x \geq 0, x \textbf{ from from } w) \longrightarrow \mathit{false}$$

Sometimes it is more convenient to specifiy the range of iteration by an interval. If *prime*? n is a predicate that returns *true* is the argument is a prime integer, the calls

$$\textbf{select}(\mathit{prime}?\ k, k \textbf{ from } 2, .., 100)$$

$$\textbf{select}(\mathit{prime}?\ k, k \textbf{ from } 3, .., 99 \textbf{ by } 2)$$

generate the set of prime integers between 1 and 100.

New sets which are obtained by applying the **separate** constructor are shown next. The first example uses the absolute value as a criterion for separation of the single set u.

$$\textbf{separate}(\mathit{abs}\ x, x \textbf{ from } u) \longrightarrow \{\{0\}, \{1, -1\}, \{4, -4\}\}$$

The next two applications show how a separation of a set into various subsets may be achieved by more complicated criteria where more than a single set may be involved.

$$\textbf{separate}(\mathit{abs}\ y, x \textbf{ from } u, y \textbf{ from } v) \longrightarrow \{\{1\}, \{0, 4\}, \{-1\}, \{3, -4\}\}$$

$$\textbf{separate}(x + y, x \textbf{ from } u, y \textbf{ from } v) \longrightarrow \{\{1, -4\}, \{0, -1, 4\}\{3\}\}$$

Finally the use of the *IterationConstructor* is shown. Assume that the two variables k and l have been assigned the values $k = 0$ and $l = 1$ first. Then two possible calls of iterate are

$$\textbf{iterate}(k := k + 1, l := 2 * l, \textbf{true}?\ k = 5) \longrightarrow 32$$

$$\textbf{iterate}(\textbf{true}?\ k = 5, k := k + 1, l := 2 * l) \longrightarrow 32$$

Their action is twofold. In either case, the value of l has been changed to 32, at the same time this is the value of the call of **iterate**. In this way a functional programming style is supported. It may be instructive to implement these examples in the language supported by the computer algebra system mentioned before. More advanced applications of these language constructs will occur in connection with the software packages described later in this section.

Design of a type system for computer algebra. In this application oriented context a type system is considered as a collection of standardized pieces of software which has been designed and implemented with the goal to enable its reuse in as many places as possible. The occurence of a certain type is grounded on the fact that it is needed in more than a single application. Theoretical questions concerning the structure of the type system are not considered here, the reader is referred to the dissertation of Weber [22] and the references given there. The proposed type system is certainly not complete. Its justification is essentially based on the fact that considerable pieces of software, mainly in the realm of differential equations, have been successfully implemented on top of it. In any case, it is probably a subsystem that will be part of any more complete system which may emerge in the course of time. A detailed description of the complete type system is far beyond the scope of the article in hand will be given elsewhere.

Name	Type	Parameter Values
Number	num	
Variable	var	
Derivative	der	
Univariate Polynomial	upoly($coef$,var)	$coef$ = num | mpoly
Multivariate Polynomial	mpoly(num,var)	
Distributive Polynomial	dpoly($coef$,var)	$coef$ = num | mpoly | mrat
Factorized Polynomial	fpoly($coef$,var)	$coef$ = integer | rn
Linear Form	lform($coef, lvar$)	$coef$ = num | upoly | mpoly | mrat | pf
		$lvar$ = var | der
Quadratic Form	qform($coef$,var),	$coef$ = num | upoly | mpoly | mrat | pf
Differential Polynomial	diffp(num,der)	
Linear Differential Form	ldf($coef$,der)	$coef$ = num | upoly | mpoly | mrat | pf
Univariate Rational Function	urat(num,var)	
Multivariate Rational Function	mrat(num,var)	
Factored Rational Function	frat(num,var)	
Partial Fraction	pf($coef$,var)	$coef$ = num | mpoly | lform
Differential Rational Function	diffrf(num,der)	

Table 11: The members of the proposed type system. Additional explanation is given in the text.

All types of the system are listed in Table 11. It is currently implemented both in *Axiom* and in *REDUCE*. A complete description must provide all the information that is necessary in order to develop software on top of it. In particular it must comprise a detailed specification of the operations for each type taking into account all possible special cases. Furthermore there should be made available as much information as possible on the cost of calling a certain operation. This topic is covered in a separate paragraph at the end of this section.

The type Number supports integers, rational numbers, algebraic numbers and modular arithmetic. The type Multivariate Polynomial realizes the completely recursive representation of a polynomial in all arguments. An object of this type may be considered as a univariate polynomial in the leading variable with coefficients which are polynomials in the remaining ones. They correspond to the standard forms of *REDUCE*. The other extreme is the type Distributive Polynomial if all variables are considered as distributive variables. However, there are also intermediate cases possible with some variables distributive and the remaining ones forming the coefficients of the type mpoly. This possibility has to be provided by an appropriate type change operation and is supported by the type system. Factorization of a polynomial is considered as a type change to the type Factorized Polynomial. The types Univariate and Multivariate Rational Function are basically quotients of upoly's and mpoly's respectively. In a Factored Rational Function the denominator is an Factorized Polynomial whereas the numerator is an upoly or mpoly respectively.

The arguments in a Differential Polynomial or Differential Rational Function may carry dependencies with them such that they may contain non-vanishing derivatives. Various kinds of homogeneous polynomials have turned out to be extremely important for applications.

This is the reason why the types Linear Form and Linear Differential Form have been chosen as exemplary cases that are described in more detail below.

In addition to the mathematical operations which are implemented for each type, there are various classes of auxiliary operations which are necessary in order to work with them.

Procedure Name & Signature	Specification of Result
$higher?(u, v : lform) : boolean$	$true$ if u is higher than v, generates a reflexive partial ordering for lform's
$reorder(u : lform) : lform$	enforces valid ordering for $lvars$'s
$neg(u : lform) : lform$	negative value of u
$lc(u : lform) : coef$	leading coefficent of u
$lnc(u : lform) : number$	leading numerical coefficient of u
$red(u : lform) : lform$	reductum of u
$mvar(u : lform) : lvar$	main variable of u
$coefficient(u : lform, v : lvar) : coef$	coefficient of v in u
$variables(u : lform) : set\ lvar$	linear variables of u
$coefficient\_variables(u : lform) set\ var$	variables occuring in coefficients
$coefficients(u : lform) : bag\ coef$	coefficients of u
$add(u, v : lform) : lform$	sum of u and v
$mult\_coef(u : lform, v : coef) : lform$	product of u and v
$new(oset\ lvar, set\ coef) : lform$	generates new $lform$ from arguments
$derive(u : lform, v : var) : lform$	derivative of u wrt v
$derive(u : lform, v : var, k : integer) : lform$	k-th derivative of u wrt v
$reduction(u, v : lform) : lform$	reduces u wrt v
$reductions(u : lform, v : set\ lform) : lform$	reduces u wrt elements of v
$auto\_reduce(u : set\ lform) : set\ lform$	autoreduced set of input u

Table 12: A Selection of operations for the type *LinearForm*. Alltogether there are more than 50 operations provided for this type. For details see the main text.

- Input- and output operations. The input of all data is performed from the user interface of some existing computer algebra system. The operations for transforming this input into the objects of the type system have to be provided for each type. The output is almost exclusively performed in Latex format.

- Type changes. Many applications require the transformation of an object of a certain type into another object of some other type without changing its mathematical meaning. For all these occurrences the type system has to provide the appropriate transformation operations.

- Random generators for creating objects of each type. They are very useful for testing, debugging and performance measurements.

Example: lform and its descendants. As an example a fairly detailed description of the type $lform(coef, lvar)$ is given in this subsection. The mathematical quantities to be modeled by the objects of this type are linear homogeneous forms. As it will turn out later on, *lform* and its descendants are the most important members of the type system as far as applications to differential equations is concerned. In order to create an object of the type *lform*, the following information has to be provided.

- The variables with respect to which linearity and homogeneity is guaranteed. They are described by the parameter *lvar*. In order to establish a unique representation, an ordering for them has to be defined. The only allowed value for the parameter *lvar* is *lvar* = var.
- The coefficients of the linear variables. The operations for *lform*'s given in Table 12 require a suitable definition of the coefficients, essentially they must have the properties of a unique factorization domain. At present the following coefficient domains are supported: *coef* = num | mpoly | dpoly | upoly | fpoly | mrat | frat | pf.

The most important descendant of *lform* is *ldf* which has also two parameters. It abbreviates *LinearDifferentialForm*. The basic difference between these two types is that for an *ldf(coef, lvar)* the second parameter is *lvar* = der. This has the consequence that in an object of type *ldf* the linear variables may also be differentiated, i. e. an *ldf* may be applied to represent a linear homogeneous differential equation. In mathematical terms the relation between these two types may be expressed by saying that a linear differential form behaves exactly like a linear form as long as no differentiations or integrations are involved. As a consequence, all operations above the single horizontal line in Table 12 may be inherited from *lform* in the implementation of *ldf* which amounts to about 50% of its code. The remaining operations have to be re-implemented in order to take into account possible dependencies of the *lvar*'s. An instructive example for this proceedings is the function *derive* which defines the first derivative wrt. to its second argument. Its implementation in *lform* is (all system dependent details as e. g. variable declarations have been omitted).

$$derive(u : lform, v : var) : lform$$

$$w := applies(derive(x, v), x\ from\ coefficients\ u)$$

$$new(variables\ u, w)$$

The derivative in the second line has the signature $derive(x : coef, v : var) : coef$, i. e. it is inherited from the coefficient type. Contrary to that, in *ldf* it is defined by

$$derive(u : ldf, v : var) : ldf$$

$$w := applies(derive(x, v), x\ from\ variables\ u)$$

$$z := new(w, coefficients\ u)$$

$$reorder\ add(z, derive(u, v))$$

Here the derivative in the second line has the signature $derive(x : der, v : var) : der$ and in the last line it is $derive(u : lform, v : var) : lform$ where an additional type change to *ldf* in the last line is implicitly assumed. It must be assured by the overall design of the software that all operations with the necessary signatures are available. This simple example shows a lot of typical features of object oriented programming which occur over and over again. Its unique goal, i. e. to avoid duplication of code by proper reuse, should be obvious from it.

In addition to those operations in *ldf* which are inherited from *lform* or which are redefined, quite a number of new operations which do not have a meaning for an *lform* are implemented. Typically these are operations which take a set of *ldf*'s as argument and transform it into a certain normal form as for example into an involutive system or into a differential Gröbner base. They have an important application in the package *SPDE* to be dicussed later on.

Another descendant of *lform* is the type *lode(der, pf)* which in turn is a subtype of *ldf*. Objects of this type model linear homogeneous ordinary differential equations. Its coefficient type is restricted to partial fractions in the independent variable. They are special *ldf*'s in the sense that only a single dependent variable in the *lvar*'s of type *der* may occur. By proper reuse of both *lform* and *ldf*, the total amount of code for implementing *lode* is cut down to about one half.

The typesystem which is discussed in this section has been realized both in *Axiom* and in *REDUCE*. Although *REDUCE* does not support any of the concepts of *OOD*, it turns out to be extremely useful to simulate these concepts by a proper organization of the software into modules. Type values are stored on global variables, its handling is supported by operations for accessing and changing them. For the module *lform* for example there exist the two functions

$$set\_types\_lform(u, v : id) : nil \qquad \text{and} \qquad reset\_types\_lform() : nil$$

with the following semantics. A call to the former function assigns the types u and v to $coef$ and $lvar$ respectively. The latter function restores the values that have been effective before this call. If these two functions are always called in pairs which could in principle be checked by the compiler, the management of the global type variables is fairly safe. A similar scheme is established for all type variables.

The *Axiom* implementation is performed along the lines described in section 4.1 utilizing all the support provided by the system. It is advantageous to collect all operations which are typical for a linear homogeneous mathematical object in a category LinearHomogeneous. A part of its implementation may be given by the following piece of *Axiom* code.

```
LinearHomogeneous(U:Ring,V:OrderedSet): Category == with
  red: $ -> $              ++ red u computes u with leading term removed
  mvar: $ -> V             ++ mvar u returns leading variable of u
  ?+?: ($,$) -> $          ++ u+v computes sum of u and v
  ?*?: ($,U) -> $          ++ u*v computes u with coefficients multiplied by v
  ?/?: ($,U) -> $          ++ u/v computes u with coefficients divided by v
  reduction: ($,$) -> $    ++ reduction(u,v) computes reduction of u wrt.
                           ++ main variable of v
  derive: ($,V) -> $       ++ derive(u,v) computes first derivative of u wrt. v
  derive: ($,V,n:Integer) -> $ ++ derive(u,v,n) computes n-th derivative of u
```

The text following the ++ in any line is a comment. The domain LFORM of type LinearHomogeneous realizes the features defined by this category a partial implementation of which might be given as follows.

```
)abbrev domain LFORM LinearForm
LinearForm(U:Ring,V:OrderedSet): Exports == Implementation where
  I ==> Integer
  Exports == LinearHomogeneous(U,V)
  Implementation == add
    Rep:List Record(coef:U, lvar:V)
    -- here follows the implementation of all exported features
```

As explained above, the type Linear Differential Form describes objects that are linear and homogeneous in a set of derivatives. Therefore the corresponding domain LDF is a member of the category LinearHomogeneous in the first place. All operations not involving dependencies are supposed to be inherited from LFORM, a few of them might be redefined and some completely new features could be included. This behavior may be achieved by the following implementation.

```
)abbrev domain LDF LinearDifferentialForm
LinearDifferentialForm(U,V): Exports == Implementation where
  U:Ring
  V:OrderedDerivative
  Exports == LinearHomogeneous(U,V) with
    invsys: Set $ -> Set $   ++ invsys u transforms u into involutive form
  Implementation == LinearForm(U,V) add
  -- here follows the implementation
```

The inclusion of category LinearHomogeneous into the category part assures that all features of LFORM are also exported by LDF. Mentioning it in the *AddDomain* branch implies that its implementations are also taken over except those that are redefined in the implementation part. Newly defined operations like *invsys* have to be mentioned explicitly in the category part. Because there is no Rep assignment it is assumed that it is taken over from *lform*.

Characteristic curves of operations. It is common practice in the engineering sciences to design sophisticated components in terms of more elementary ones. A fundamental prerequisite for this proceeding is a detailed and quantitative description of the latter, comprising all attributes that may be relevant for the application one has in mind. A typical example of this procedure is the use of transistors in an electrical circuit. They are described by a bulk of characteristics relating the input and output parameters in numerous ways. Starting from them, the most suitable components are chosen and the circuit with the desired functionality may be realized in terms of them.

Although the development of algorithms is a process of comparable complexity, the methods for designing and implementing them are far less developed. To a considerable extent they are based on heuristics or even trial and error. On second thoughts it is realized that the

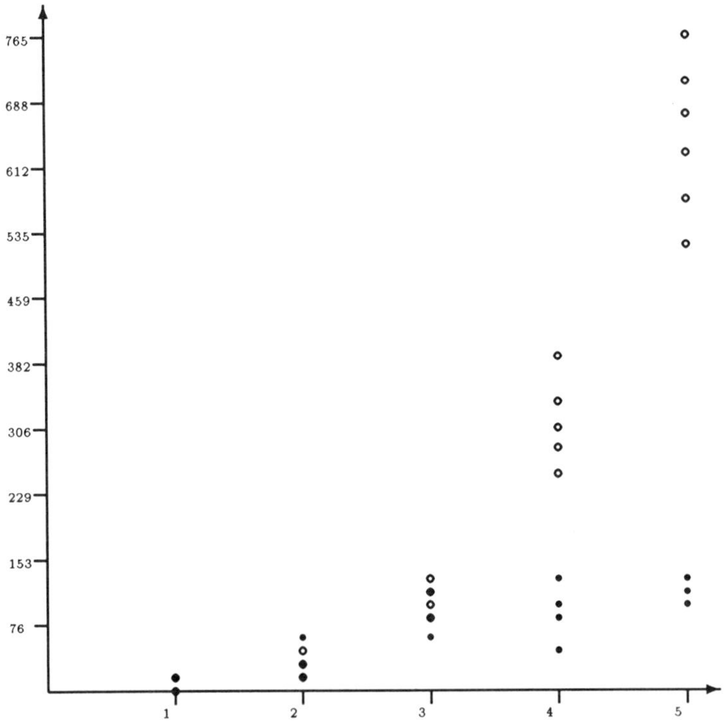

Figure 8: Modular and non-modular gcd computations are compared. Details are explaind in the text.

reason is a lack of quantitative information about the components to be used for implementing high-level operations. Therefore it is advocated that the specification of any operation in the type system should be supplemented by a set of *characteristic curves* for various regions in the parameter space of the accepted inputs. These curves should describe the behavior for input distributions that may occur in applications. In particular they should also contain information on the worst case behavior. This latter information is required in order to understand and possibly avoid the highly undesired feature that a piece of software never returns a result upon submitting certain inputs. In this manor it should be possible to design high level algorithms based on this type system in a systematic way similar as it is established in the engineering sciences such that it deserves the name *algorithm engineering*.

In order to illustrate this process, a few examples for characteristic curves will be given. All computations have been performed on a SUN Sparc station in *REDUCE* 3.5. The ordinate indicates always the computing time in milliseconds. In Figure 8 the abscissa gives the total degree of the two-variate input polynomials. These polynomials are dense, its coefficients are integers with not more than three decimal digits. The open circles correspond to a polynomial remainder sequence calculation, the heavy dots to a modular EZGCD algorithm. This plot shows clearly that the latter algorithm is far better for a total degree of four or more whereby the relative benefit increases strongly with the degree. For degrees less than four there is no significant difference. Both algorithms show a smooth behavior in the sense that for a fixed total degree the computing time differs not more than about 20%.

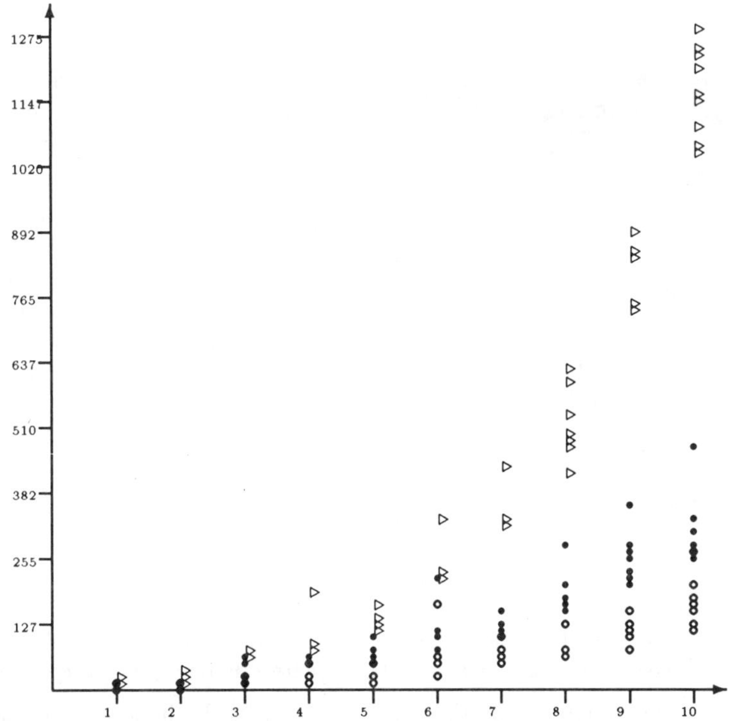

Figure 9: This plot shows the factorization of polynomials over the integers. Details are explained in the text.

In Figure 9 several factorizations are compared. The open circles, heavy dots and triangles correspond to univariate, bivariate and trivariate polynomials respecetively. All polynomials are dense with coefficents not larger than 10^3 in absolute value. This plot gives a complete quantitative description of all possible cases in this region of the parameter space for the input. Based on this information, the behavior of any other algorithm which uses this factorization may be quantitavely predicted within fairly narrow error margins.

For many algorithms there is almost an infinite number of conceivable combinations of input parameters. A systematic investigation should find out the significant cases, pointing out characteristic differences and provide this information concisely as part of the description of the type system. In addition it is highly desirable to include into this description as much information as possible that follows from the theoretical analysis and to compare it with the empirical data.

5.2 SPDE: Symmetries of Partial Differential Equations

Symmetry analysis of differential equations has been introduced by Sophus Lie in the second half of the last century in order to obtain an algorithmic way for solving nonlinear differential equations. Yet it is rarely applied for that purpose and most textbooks do not even mention it. The reason is probably the enormous amount of calculations which is usually involved in obtaining the symmetry group of a given differential equation. Therefore the package *SPDE* which abbreviates **S**ymmetries of **P**artial **D**ifferential **E**quations has been developed. It returns the complete symmetry group for a wide class of differential equations automatically. Its implementation along the lines explained in the preceding sections will now be described.

Mathematical background. As far as the mathematics is concerned, only those results are given which are necessary for understanding the subsequent sections. Details and many examples including a fairly complete list of references may be found in the review article on that subject by the author [25]. The most general case of a system of partial differential equations for an arbitrary number m of unknown functions u^α which may depend on n arguments x_i is considered. These variables are collectively denoted by $u = (u^1, \ldots, u^m)$ and $x = (x_1, \ldots, x_n)$ respectively. The system of N differential equations is written in the form

$$\omega_\nu(x_i, u^\alpha, u_i^\alpha, u_{ij}^\alpha, \ldots, u_{i_1,\ldots,i_k}^\alpha) = 0 \tag{1}$$

for $\nu = 1, \ldots, N$ where the notation

$$u_{i_1 \ldots i_n}^\alpha = \frac{\partial^{i_1 + \ldots i_n} u^\alpha}{\partial x_1^{i_1} \ldots \partial x_n^{i_n}}$$

for derivatives has been used. Furthermore it is assumed that the equations (1) are polynomial in all arguments. For $m = n = N = 1$ a single ordinary differential equation is obtained.

To formulate the condition for the invariance of (1), the infinitesimal generator U is defined by

$$U = \xi_i \frac{\partial}{\partial x_i} + \eta^\alpha \frac{\partial}{\partial u^\alpha} \tag{2}$$

where ξ_i and η^α may depend on all dependent and independent variables. Summation over twice occuring indices is always assumed. Greek indices run from 1 to m and latin indices from 1 to n. The k-th prolongation of U is defined as

$$U^{(k)} = U + \zeta_i^\alpha \frac{\partial}{\partial u_i^\alpha} + \ldots + \zeta_{i_1,\ldots,i_k}^\alpha \frac{\partial}{\partial u_{i_1,\ldots,i_k}^\alpha}. \tag{3}$$

where the functions $\zeta_{i_1,\ldots,i_k}^\alpha$ describe the transformation of partial derivatives of order k. The $\zeta's$ satisfy the recursion relations

$$\zeta_i^\alpha = D_i(\eta^\alpha) - u_s^\alpha D_i(\xi_s) \tag{4}$$

and

$$\zeta_{i_1,\ldots,i_k}^\alpha = D_{i_k}(\zeta_{i_1,\ldots,i_{k-1}}^\alpha) - u_{i_1,\ldots,i_{k-1},s}^\alpha D_{i_k}(\xi_s). \tag{5}$$

where

$$D_i = \frac{\partial}{\partial x_i} + u_i^\alpha \frac{\partial}{\partial u^\alpha} + u_{ki}^\alpha \frac{\partial}{\partial u_k^\alpha} + u_{kli}^\alpha \frac{\partial}{\partial u_{kl}^\alpha} \cdots. \tag{6}$$

is the operator of total derivation with respect to x_i. The system of differential equations (1) is invariant under the transformations of a one-parameter group with the infinitesimal generator (2) if the ξ's and η's are determined from the conditions

$$U^{(k)}\omega_\nu = 0 \quad \textit{when} \quad \textit{all} \quad \omega_\nu = 0. \tag{7}$$

Under the constraints for the ω_ν which have been mentioned above, the left hand side of (7) is a polynomial in all its arguments. Because the derivatives of the $u's$ do not occur as arguments of the ξ's and the $\eta's$, it has to be decomposed with respect to these derivatives and the coefficients are equated to zero. The resulting set of equations is called the *determining system* for the symmetries. Its general solution determines the full symmetry group of (1). More precisely, to each constant occuring in the general solution there corresponds a symmetry generator of a finite Lie group, each unknown function determines an infinite parameter group.

Example: The Korteweg deVries equation. In order to illustrate the procedure for determining the Lie symmetries, an example will be worked out in detail. To this end, consider the well known Kortweg deVries equation which has become famous for its soliton solutions. It is simple enough to be presented on less than a page, nevertheless it shows all steps that are typical for the general case. In the notation introduced above it reads

$$u_{2,2,2} + u_2 u + u_1 = 0 \tag{8}$$

where $u \equiv u(x_1, x_2)$, $u_1 \equiv \partial u/\partial x_1$ etc. Applying the third prolongation on the left hand side and substituting the equation itself, the determining system is obtained in the form

$$\begin{gathered}
\xi_{1,u,u} = 0, \ \xi_{1,u,x_2} = 0, \ \xi_{1,u,u,u} = 0, \ \xi_{1,u,u,x_2} = 0, \ \xi_{2,u} - \xi_{1,u,x_2,x_2} = 0 \\
\xi_{2,x_2} - \frac{1}{3}\xi_{1,x_1} - \frac{1}{3}\xi_{1,x_2,x_2,x_2} - \frac{1}{3}u\xi_{1,x_2} = 0, \ \xi_{2,u} = 0, \ \xi_{2,u,u} = 0 \\
\eta_{u,u} - 3\xi_{2,u,x_2} = 0, \ \xi_{1,u} = 0, \ \eta_{u,x_2} - \xi_{2,x_2,x_2} = 0, \ \xi_{2,u,u,u} = 0 \\
\eta_{u,u,u} - 3\xi_{2,u,u,x_2} = 0, \ \xi_{1,u,u} = 0, \ \eta_{u,u,x_2} - \xi_{2,u,x_2,x_2} + u\xi_{2,u} = 0 \\
\xi_{1,u} = 0, \ \xi_{1,u,x_2} = 0, \ \eta_{u,x_2,x_2} + \frac{1}{3}\eta - \frac{1}{3}\xi_{2,x_1} - \frac{1}{3}\xi_{2,x_2,x_2,x_2} + \frac{2}{3}u\xi_{2,x_2} = 0 \\
\xi_{1,x_2} = 0, \ \xi_{1,x_2,x_2} = 0, \ \eta_{x_1} + \eta_{x_2,x_2,x_2} + u\eta_{x_2} = 0
\end{gathered} \tag{9}$$

The first step is to transform it into a differential Gröbner base with the result

$$\begin{gathered}
\xi_{1,u} = 0, \ \xi_{1,x_2} = 0, \ \xi_{1,x_1,x_1} = 0, \ \xi_{2,u} = 0 \\
\xi_{2,x_2} - \frac{1}{3}\xi_{1,x_1} = 0, \ \xi_{2,x_1,x_1} = 0, \ \eta - \xi_{2,x_1} + \frac{2}{3}u\xi_{1,x_1} = 0
\end{gathered} \tag{10}$$

At this point, all functions depend on all variables x_1, x_2 and u. The term ordering applied is a lexicographic ordering with $\eta > \xi_2 > \xi_1$ and $x_1 > x_2 > u$. The general solution is obtained as

$$\xi_1 = c_1 - 3c_3 x_1, \quad \xi_2 = c_2 x_1 - c_4 x_2 - c_6, \quad \eta = c_2 + 2c_3 u \tag{11}$$

with $c_1, \ldots, c_4$ constants. Each of them corresponds to a symmetry generator with the final result

$$U_1 = \frac{\partial}{\partial x_1}, \quad U_2 = \frac{\partial}{\partial x_2}, \quad U_3 = x_1 \frac{\partial}{\partial x_2} + \frac{\partial}{\partial u}, \quad U_4 = x_1 \frac{\partial}{\partial x_1} + \frac{1}{3} x_2 \frac{\partial}{\partial x_2} - \frac{2}{3} u \frac{\partial}{\partial u} \qquad (12)$$

and the non-vanishing commutators

$$[U_1, U_3] = U_2, \quad [U_1, U_4] = U_1, \quad [U_2, U_4] = \frac{1}{3} U_2, \quad [U_3, U_4] = -\frac{2}{3} U_2 \qquad (13)$$

Identifying the objects and classes. The first step towards an implementation is to identify the mathematical objects that occur in the problem and to group them together according to some common attributes. This process is a very important first step in the design of a piece of computer algebra software. A favourable choice at this point may very well determine the quality and overall performance of the software to be developed. In the first place there are various kinds of variables and functions.

- The independent and dependent variables $x = (x_1, x_2, \ldots, x_n)$ and $u = (u^1, u^2, \ldots, u^m)$ respectively and the derivatives of the latter. Furthermore, in practical problems there may occur constant parameters $a = (a_1, a_2, \ldots)$. A common feature of these quantities is that they occur in the given differential equations.
- The functions ξ_i and η^α determine the symmetry generator. They occur as the unknown functions of the determining system. In the course of the solution procedure, there may appear additional functions c_i. Their distinguishing common feature is that the equations of the determining system are linear and homogeneous in these functions and they may depend on all dependent and independent variables.
- The partial derivative operators $\partial/\partial x_i$ and $\partial/\partial u^\alpha$ occur in the final answer representing the symmetry generators.

Out of these variables various kinds of mathematical objects are formed. The notation ∂f means any meaningful derivative of the function f, including f itself.

- The given differential equations. Under the assumptions mentioned at the beginning of this section they are from $\mathbf{Q}[x, u, a]$.
- The equations of the determining system are linear and homogeneous in the ξ's, η's, c's and its derivatives, i. e. they are from $\mathbf{Q}(x, u, a)[\partial\xi, \partial\eta, \partial c]$.
- The infinitesimal generators of a finite symmetry group are linear and homogeneous in the partial derivative operators $\partial/\partial x_i$ and $\partial/\partial u^\alpha$ with rational function coefficients depending on all variables occuring in the differential equations including the parameters, i. e. they are from $\mathbf{Q}(x, u, a)[\partial/\partial x_i, \partial/\partial u^\alpha]$.
- The infinitesimal generators of an infinite symmetry group are linear and homogeneous in the partial derivative operators $\partial/\partial x_i$ and $\partial/\partial u^\alpha$. Its coefficients may be considered as equations of the determining system which have not been explicitly solved. This implies the structure $\mathbf{Q}(x, u, a)[\partial\xi, \partial\eta, \partial c][\partial/\partial x_i, \partial/\partial u^\alpha]$.

Mathematical Domain	Object for Implementation
$\mathbf{Q}[x,u,a]$	$diffp(rn, der)$
$\mathbf{Q}(x,u,a)[\partial\xi,\partial\eta,\partial c]$	$ldf(mrat(rn, var), der)$
$\mathbf{Q}(x,u,a)[\frac{\partial}{\partial x_i},\frac{\partial}{\partial u^\alpha}]$	$lform(mrat(rn, var), var)$
$\mathbf{Q}(x,u,a)[\partial\xi,\partial\eta,\partial c]\,[\frac{\partial}{\partial x_i},\frac{\partial}{\partial u^\alpha}]$	$lform(lform(mrat(rn, var), var), var)$

Table 13: Datatypes assignments for the package *SPDE*. Details are explained in the text.

Until now the language applied for describing the problem at hand has been purely mathematical, there has been no mentioned of computer algebra software whatsoever. The next and most important step in the design of the package *SPDE* is to match these mathematical quantities with the objects provided by the type system. The answer is given in Table 13. As it has been mantioned before, the most important datatypes occuring in the design of *SPDE* are *lform* and its descendants, they comprise about 80 % of the total code. As a matter of fact, also the type *lode* plays an important role because frequently equations of the determining system may be coerced to this type and utilize the solution facilities implemented there. The choice of *diffp* insted of *mpoly* has the following reason. In some applications there occur unspecified functions of the indepedent variables in the given differential equations, they are modeled as parameters carrying some dependencies with them. In order to take these dependencies into account it suffices to pass over from *mpoly* to *diffp* as it has been explained in section 5.1 where the type system was explained. This is good example of how by proper choice of a datatype a considerable effect is achieved.

Getting the algorithms to work. After the major decisions concerning the overall design of a software package have been made, the algorithms which realize the desired functionality have to be coded. On top of a well designed type system and by means of properly chosen data types for the problem at issue, the expense for this process will be minimized. It will be dicussed now in some detail how it works for the package *SPDE*. To this end the full process of symmetry analysis is decomposed in various parts, guided by the example of the Korteweg deVries equation given above. The major steps are as follows.

- Generate the determining system from the given differential equations and transform it into a differential Gröbner base. The latter step is important for two reasons. In the first place it simplifies the determining system and thereby faciliates the solution procedure. Secondly from a Gröbner base the size of the symmetry group may easily be read off [26] before the solution algorithm is entered (compare eqs. (9) and (10)).
- Solve the determining system. In order to obtain the full symmetry group, it has to be assured that the general solution is obtained. For large classes of equations it can be shown that this process is a decision procedure (compare eq. (11)).
- Construct the infinitesimal generators of the symmetry group from the general solution and compute the commutator table (compare eq. (12)).

Generating the determining system from the differential equations requires essentially differentiations and substitutions. There is a function

cresys(set diffp) : set ldf

provided for that purpose that performs the various steps, mainly by calling operations from *diffp* and *ldf*. The algorithm for transforming the result into a differential Gröbner base is completely implemented in *ldf* because it is not specific to symmetry analysis.

The largest portion of the code in *SPDE* realizes the solution algorithm for the determining system. A few examples from this part of *SPDE* will be used to demonstrate the application of the language described in section 5.1. An important simplification step is the *separation* of an equation of the determining system. It may be applied if the dependence on one or more variables has become completely explicit. An example of this procedure is given by the equation

$$(x^2 + y^2 + z^2)f_{yx} + z^2 f_x + (x + z^2)g_{yy} + yg_y = 0$$

where $f \equiv f(x, y)$ and $g \equiv g(y)$. Collecting coefficients of the variable z, it may obviously be replaced by the two simpler equations

$$f_{yx} + f_x + g_{yy} = 0, \qquad (x^2 + y^2)f_{yx} + xg_{yy} + yg_y = 0$$

In order to implement this process as a simplification step for a determining system, it is broken up into the following more elementary operations. In the first place a function *separation_variables* with the signature

separation_variables(u : ldf(mpoly, der)) : set var

is needed that returns the variables with respect to which a separation of an *ldf* may be performed. This function comes from *ldf*. Secondly the actual breakup has to be achieved for a given *ldf* and a given set of separation variables. The corresponding function *separation* with the signature

separation(u : lform(mpoly, var)) : set lform

is *not* from *ldf* but from *lform* because this is an operation where no dependency or differentiation is involved. It is defined for any polynomial, in fact the implementation in *lform* is to a large part inherited from *mpoly* and *dpoly*. This is an example of the extremely important principle that any operation is moved up in the inheritance hierarchy to the module with the minimal number of features that are necessary for it to be meaningful. In this manor its reuse in as many occasions as possible is assured. If now to the variable u of type *set ldf* there has been assigned a determining system, all possible separations and the subsequent transformation into a Gröbner base are achieved by the two lines of code in the package *SPDE*

v:=applies(separation_variables x,x from u)

z:=invsys concats applies(separation(x,y),x from u,y from v)

Another important simplification occurs if there are equations of the form $f_x = 0$, they may be used to remove the corresponding dependency from f. Let again u have a determining system as its value. Then the following piece of code performs this kind of simplification.

u0:=select(red x=0 and order x=1,x from u)

v:=applies(mvar x,x from u)

vd:=applies(function x,x from v)

vi:=applies(derivatives x,x from v)

applies(remove_dependencies(x,y),x from vd,y from vi)

u:=invsys applies(simplify x,x from u)

The first line picks up equations of the desired form from u where *red* is from *lform* (see Table 12) and *order* is from *ldf*.The second line assigns the set of first derivatives to the variable v where *mvar* is again from *lform*, *vd* and *vi* collect the function and the variable wrt. which it has been derived, the respective operations *function* and *derivatives* are from *der*. The next line removes all possible dependencies. Finally the elements from u are simplified and u is transformed into a Gröbner base.

There are two more simplification steps for the determining system, i. e. substitutions and solving genuine ordinary differential equations. The former process is essentially an operation inherited from *lform*, that latter is received from the datatype *lode* where also the solution algorithms are implemented.

After the determining system has been solved, two more actions are required. The symmetry generators have to be constructed from the general solution, and the commutator table for the generators of the finite Lie group has to be obtained. These operations are mainly implemented in *SPDE* because they are genuine features of the symmetry analysis. Basically they are achieved by collecting the contributions from the various constants and functions in the general solution and attaching the proper partial derivative operators wrt. the dependent and the independent variables to them. Most operations for these procedures may be inherited from the type system. Applying the language constructs from above this is achieved with less than one hundred lines of code.

Description of the user interface. The interaction of a user with a given piece of software occurs by means of the user interface. As a matter of fact, a well designed and implemented software package should not require any additional knowledge beyond the description of this interface. In many cases, a user does not even want to know the internal working of the software he needs in his applications. The most important user interface function *symmetries* of the package *SPDE* has the signature

symmetries mpoly or *symmetries(set mpoly)*

Its single argument represents a single differential equation or a system of such equation respectively. In general, the symmetry generators of the goup of Lie symmetries will be returned. If there occur unspecified functions in the argument, only the determining system will be displayed because in these cases a solution algorithm cannot be designed. A complete session producing the answer for the above mentioned Korteweg deVries equation proceeds as follows.

```
depend u,x,t; symmetries df(u,x,t)+u*df(u,x)+df(u,x,3);
```

In this example *REDUCE* syntax has been applied. The *depend* statement at the beginning is necessary in order to communicate to the system that u is a function of both x and t. After about 10 seconds the system will return the answer (12), (13).

As a second example consider the following system by Noor [27] corresponding to the large deflection analysis of anisotropic shallow shells.

$$\frac{1}{2}w_x^2 + k_1 w + u_x = 0, \qquad \frac{1}{2}w_y^2 + k_2 w + v_y = 0, \qquad w_x w_y + 2k_{12} w + v_x + u_y = 0$$
$$w_{xx} = w_{yy} = w_{xy} = 0 \tag{14}$$

where u, v, w depend both on x and y. Submitting the input

```
depend {u,v,w},{x,y};

noor:={1/2*df(w,x)^2+k1 w+df(u,x)=0 ,1/2*df(w,y)^2+k2 w+df(v,y)=0,

       df(w,x)*df(w,y)+2*k12*w+df(v,x)+df(u,y)=0,

       df(w,x,2)=0, df(w,y,2)=0, df(w,x,y)=0};

symmetries noor;
```

the following output is generated.

$$U_1 = \frac{\partial}{\partial x}, \quad U_2 = \frac{\partial}{\partial y}, \quad U_3 = \frac{\partial}{\partial u}, \quad U_4 = \frac{\partial}{\partial v}, \quad U_5 = y\frac{\partial}{\partial u} - x\frac{\partial}{\partial v}$$

$$U_6 = (2k_{12}^2 y^2 + 2k_{12}k_1 xy - \frac{1}{2}k_1 k_2 y^2 + \frac{1}{2}k_1^2 x^2 + k_1 w)\frac{\partial}{\partial u}$$
$$+(k_{12}k_2 y^2 + k_1 k_2 xy + 2k_{12} w)\frac{\partial}{\partial v} + (2k_{12} y + k_1 x)\frac{\partial}{\partial w}$$

$$U_7 = (2k_{12} y + k_1 x)\frac{\partial}{\partial u} + k_2 y\frac{\partial}{\partial v} + \frac{\partial}{\partial w}$$

$$U_8 = (k_{12}^2 + k_1 xy)\frac{\partial}{\partial u} + (\frac{1}{2}k_2 y^2 - \frac{1}{2}x^2 + w)\frac{\partial}{\partial v} + y\frac{\partial}{\partial w}$$

with the non-vanishing commutators

$$[U_1, U_5] = -U_4, \quad [U_1, U_6] = k_1 U_7, \quad [U_1, U_7] = k_1 U_3, \quad [U_1, U_8] = k_1 U_5$$

$$[U_2, U_5] = U_3, \quad [U_2, U_6] = 2k_{12}U_7 - k_1 k_2 U_5, \quad [U_2, U_7] = 2k_{12}U_3 + k_2 U_4, \quad [U_2, U_8] = U_7$$

$$[U_6, U_7] = -k_1 U_3 - 2k_{12}U_4, \quad [U_6, U_8] = -k_1 U_5, \quad [U_7, U_8] = U_4$$

5.3 FINDS: First Integrals of Dynamical Systems.

In various areas of applied mathematics there occur autonomouos systems of ordinary differential equations, they describe for example chemical reactions, population growth, the behavior of lasers and many other processes. One of the most prominent features governing the solution set of autonomous systems is the highly irregular or *chaotic* behavior which may occur in some regions of the parameter space as opposed to a regular behavior which may be characteristic for some other region. It is one of the basic problems connected with such systems to characterize these regions. There is no general answer to this problem. The situation is completely different however if first integrals can be found.

Mathematical background. The following notation will be used throughout this section. Let $\mathbf{x} = (x_1, ..., x_n)$ be n functions depending on the variable t and let

$$\Omega_i(t, \mathbf{x}) \equiv \dot{x}_i - \omega_i(\mathbf{x}, \mathbf{c}) = 0 \qquad (i = 1, ..., n) \tag{15}$$

be a system of n *autonomous* differential equations, i. e. the independent variable t does not occur explicitly in the ω's; n is called the order because under rather weak constraints the system (15) is equivalent to a single ordinary equation of order n. In addition they may contain one or more parameters c_k which are denoted collectively as $\mathbf{c} = (c_1, c_2, \ldots)$. It is always assumed that the ω_i are polynomial in all arguments with rational number coefficients, i. e. $\omega_i \in \mathbf{Q}[\mathbf{x}, \mathbf{c}]$. By definition a function f is a *first integral* for (15) if its total derivative with respect to time vanishes, i.e. if it satisfies

$$\frac{df}{dt} = 0 \tag{16}$$

under the constraints (15). If f does not depend on t explicitly this condition may be written as

$$\sum_{i=1}^{n} \omega_i \frac{\partial f(\mathbf{x})}{\partial x_i} = 0 \tag{17}$$

and the function $f(\mathbf{x})$ is called a *time independent first integral.* There may exist also time dependent first integrals with exponential time dependence of the form

$$f(t, \mathbf{x}) = e^{\alpha t} f(\mathbf{x}). \tag{18}$$

In this case condition (16) leads to the constraints

$$\alpha f(\mathbf{x}) + \sum_{i=1} \omega_i \frac{\partial f(\mathbf{x})}{\partial x_i} = 0 \tag{19}$$

The problem of finding first integrals of the dynamical system (15) may now be formulated as follows. Find those constraints for the parameters $\mathbf{c}$ and α such that a first integral of a certain type is allowed.

Example: The Lorenz system. It is instructive to pursue the various steps for finding first integrals by means of an example. To this end consider the well known Lorenz system which is written in the form

$$\dot{x} = -ax + ay, \quad \dot{y} = bx - xz - y, \quad \dot{z} = -cz + xy$$

Here x, y and z are functions of the time t, the dot means differentiation wrt. to time; a, b and c are time-independent parameters. In order to find the quasipolynomial integrals up to second order, the ansatz

$$f(x, y, z, t) = z^2 a_9 + zya_8 + zxa_7 + y^2 a_6 + yxa_5 + x^2 a_4 + za_3 + ya_2 + xa_1 + a_0)e^{\alpha}t$$

is made. It leads to the determining system

$$a_7(2c-\alpha)=0,\;\; a_2(c-\alpha)=0,\;\; a_9(\alpha-2)=0$$
$$a_4(a-\alpha)=0,\;\; a_7-a_9=0,\;\; 2c_2a_9+2aa_4+a_2=0$$

The first solution

$$a_1=a_2=a_5=a_6=a_8=a_9=0,\;\; a_3=a_7=-a_4$$
$$a=c=1,\;\; \alpha=2$$

is one-dimensional in the space of the parameters a, b, c and α. The second solution

$$a_1=a_3=a_5=a_6=a_8=a_9=0,\;\; a_2=-\alpha a_4$$
$$a=\tfrac{1}{2}\alpha,\;\; c=b$$

is two-dimensional. They yield the two first integrals

$$\alpha-c=0,\;\; a-\frac{1}{2}c=0: \qquad I(t,\mathbf{x})=(cz-x^2)e^{ct} \tag{20}$$

$$c-1=0,\;\; \alpha-2=0,\;\; a-1=0: \qquad I(t,\mathbf{x})=(-bx^2+y^2+z^2)e^{2t} \tag{21}$$

At the left hand side the constraints for the parameters are given such that the corresponding first integral at the right hand side does exist.

Implementation aspects. The various steps from the description of a problem in mathematical terms to a piece of software with a well defined functionality proceed similarly as it has been explained for the package *SPDE*. Therefore the discussion will be much shorter in this case. There occur basically two kinds of mathematical quantities, i. e. the equations of the dynamical system and the equations of the determining system. Their representation in terms of datatypes may be seen from the following table.

Equation from	Mathematical Domain	Data Type
dynamical system	$\mathbf{Q}[c, x, \dot{x}]$	$diffp(rn, var)$
determining system	$\mathbf{Q}[c][a]$	$lform(mpoly(rn, var), var)$

The equations of the determining system are linear and homogeneous in the variables originating from the ansatz $a = (a_1, a_2, \ldots)$. This is the outermost layer of variables in the type *lform* in the last line. The variables occuring as arguments of the *mpoly*'s represent the parameters c, they form the coefficients of the a's. Similar as in the symmetry analysis, the complete process of finding first integrals decomposes into three major steps.

- Generate the determining system from the given differential equations. This process involves essentially differentiation, substitution and separation wrt. to certain kinds of variables. Most operations for this step are available in *diffp* and *mpoly*.

- Solve the determining system. This part is far more extensive than the rest. However, due to the special structure of the equations of the determining system, a large amount the operations needed for implementing the solution algorithms are inherited from *lform*. A special kind of complication arises from the fact that due to numerous factorizations in the course of the solution procedure a given problem splits into a large number of smaller ones. A sophisticated bookkeeping process has to be stablished in order to insure that a complete set of solutions is finally obtained.

- Analyse the solutions obtained in the preceeding step and construct the first integrals from them. In order to produce a meaningful result, a minimal set of first integrals avoiding any duplications or dependencies has to be generated.

Following these guiding principles, less than one thousand lines of code are sufficient to implement the package *FINDS* on top of the type system described above. For the customary user the interaction with this package occurs though the user interface function

first_integrals(u:set diffp,n:integer,v:id)

with the following semantics. The first argument u passes the equations of the dynamical system to this function call, the second argument n fixes the degree up to which first integrals are searched for, the last argument v is a label that currently may assume the two values $v =$ poly | qpoly. It specifies the type of first integrals to be searched for and is currently limited to these two choices. The following session will produce the quasipolynomials first integrals of order two for the Lorenz system

```
depend {x,y,z},t;

lorenz:={df(x,t)=a*(y-x),df(y,t)=b*x-x*z-y,df(z,t)=-c*x+x*y};

first_integrals(lorenz,2,qpoly);
```

The generalization to integrals of higher order is obvious.

6 Summary and Conclusion

The central subject of this article have been two basic questions: How to make the process of developing computer algebra software on a large scale (10^4 to 10^5 lines of code or more) more efficient and how to improve the quality of the result. Taking procedures from well established engineering sciences as a guide, two fundamental principles turned out to be of overwhelming importance: Modularization and limitation of growth through reuse. Important means for achieving these goals turned out to be the concept of an abstract data type and the principles of object-oriented design. It is advocated to install an additional abstraction level between the mathematics and the machine in order to render it possible to develop (computer algebra) system independent mathematical software. Basic constitutents of this level are a type system and a high-level language.

The ultimate criterion for rating the success of software development is the scope of functionality that can finally be reached before control is eventually lost. This question is important to such a degree because at present only a tiny fraction of mathematical problems that could possibly be handled by computer algebra tools has actually been tackled. In the realm of differential equations for example it is estimated that of all suitable problems only a few percent have actually been dealt with. In the engineering sciences it is probably not much different.

In order to advance from today's state of affairs high-quality computer algebra software should be developed for as many applications as possible. Very often the experience that is gained in this process and in particular the feedback from the users may be put into a better and more powerful piece of software that replaces the original one. The current versions of the packages $SPDE$ and $FINDS$ which have been described in section 5 originate from such a process.

In addition to the mathematical and algorithmic capabilities, it is very important to increase the acceptance of these new tools by providing a user interface as convenient as possible. As far as the input is concerned, the *intelligent sheet of paper* that has been developed under the guidance of Prof. Hotz at the University of Saarbrücken [28] should be mentioned in this context. It renders it possible to submit input for computer algebra software in truly two-dimensional handwritten form from a notepad.

Acknowledgement. A critical reading of some parts of the manuscript by Jürgen Weiß is gratefully acknowledged.

References

[1] E. Kreuzer, G. Leister, *Programmsystem NEWEUL'90*, Manual, University of Stuttgart, Institut B for Mechanics, AN-24, 1990.

[2] A. C. Hearn, *Reduce User's Manual*, part of Reduce.

[3] M. MacCallum, F. Wright, *Algebraic Computing with Reduce*, Clarendon Press, Oxford, 1991

[4] F. Schwarz, *A Guided Tour Through REDUCE.*

[5] B. Buchberger, G. E. Collins, R. Loos, Eds., *Computer Algebra, Symbolic and Algebraic Computation*, Computing Supplementum 4, Springer, 1982.

[6] J. D. Davenport, Y. Siret, E. Tournier, *Computer Algebra*, Academic Press, 1988.

[7] K. O. Geddes, S. R. Czapor, G. Labahn, *Algorithms for Computer Algebra*, Kluwer Academic Publishers, 1993.

[8] R. Zippel, *Computer Algebra Algorithms*, Kluwer Academic Publishers, 1993.

[9] R. D. Jenks, B. Sutor, *Axiom*, Springer, 1992.

[10] R. S. Pressman, *Software Engineering*, McGraw-Hill, 1992.

[11] G. Booch, *Object Oriented Design*, Benjamin/Cummings Publishing, 1991.

[12] P. Coad, E. Yourdon, *Object-Oriented Analysis* and *Object-Oriented Design*, both published by Yourdon Press, Englewood Cliffs, 1991.

[13] B. Meyer, *Object-oriented Software Construction*, Prentice Hall, 1988.

[14] B. Liskov and J. Guttag, *Abstraction and Specification in Program Development*, The MIT Press, Cambridge, 1986.

[15] L. Cardelli, P. Wegner, *On Understanding Types, Data Abstraction, and Polymorphism*, Computing Surveys **17**, 471-522 (1985).

[16] O. J. Dahl, K. Nygaard, *SIMULA 67 Common Base Proposal*, Norwegian Computing Center Document, Oslo, 1967.

[17] A. Goldberg, D. Robson, *Smalltalk-80*, Addison Wesley, 1983.

[18] N. Wirth, *On the Composition of Well-Structured Programs*, Computing Surveys **6**, 247-259 (1974).

[19] T. Budd, *Object-Oriented Programming*, Addison-Wesley, 1991.

[20] J. F. Fedoraro, *The Design of a Language for Algberaic Computation Systems*, Thesis, Berkeley, 1983.

[21] N. J. Lehmann, *Probleme und Bedeutung der Entwicklung von Programmiersprachen*, Elektronische Informationsverarbeitung und Kybernetik **11**, 200-206 (1975).

[22] A. Weber, *Structuring the Type System of a Computer Algebra System*, Dissertation, Universität Tübingen, 1992.

[23] J. Schwartz, R. B. K. Dewar, R. Dubinksy, E. Schonberg, *Programming with Sets: A Introduction to SETL*, Springer, New York, 1986.

[24] *Scratchpad II User's Guide*, Computer Algebra Group at IBM Yorktown Heights.

[25] F. Schwarz, *Symmetries of Differential Equations: From Sophus Lie to Computer Algebra*, SIAM Review **30**, 450-481 (1988).

[26] F. Schwarz, *An Algorithm for Determining the Size of Symmetry Groups*, Computing **49**, 95-115 (1992).

[27] A. K. Noor, *Computerized Symbolic Manipulation in Structural Mechanics*, these Proceedings.

[28] G. Hotz, *Optical Formula Recognizer*, System Manual, University of Saarbrücken, 1990.

COMPUTERIZED SYMBOLIC MANIPULATION IN STRUCTURAL MECHANICS

A.K. Noor
University of Virginia, Hampton, VA, USA

ABSTRACT
Status and some selected applications of computerized symbolic manipulation to structural mechanics problems are summarized. The applications discussed include: a) deriving the governing nonlinear differential equations for shallow shells; b) determination of the stationary conditions for variational functionals; c) application of Rayleigh-Ritz and Bubnov-Galerkin techniques; and d) generation of characteristic arrays for finite elements. The major advantages of using computerized symbolic manipulation in each of these applications are outlined. Other reported applications are discussed and future directions for research are outlined.

1. INTRODUCTION

Although the first documented use of the computer to perform symbolic manipulations dates back to 1953 (Refs. [1] and [2]), it was only in the last two decades that symbolic manipulation has emerged as a new discipline combining computer science and mathematics. To date the applications of symbolic computations range over the entire scope of mathematics, as well as the different branches of science and engineering. Some of these applications are described in Refs. [3], [4] and [5].

Notable advances in symbolic computation continue to take place on a broad front. The new advances are manifested by the development of fast algorithms, as well as of versatile and powerful interactive software systems for performing a wide variety of symbolic computations. The software systems incorporate mathematical knowledge and may be viewed as expert systems. They have improved the productivity of scientists and engineers. A list of some of the general purpose and specialized symbolic manipulation software systems is given in Tables 1 and 2. The Macsyma and Reduce systems date back to the 1960s. Some attempts have been made to review, evaluate and compare the different symbolic manipulation systems (see, for example, Refs. [6] to [13]). Based on the available information, no single symbolic manipulation system is likely to satisfy *all* the needs for structural mechanics applications.

Most of the current symbolic manipulation systems have, in addition to their symbolic computation facilities: a comprehensive library of functions; numerical and visualization capabilities; debugging and monitoring facilities; Fortran and C Code generation facilities; and worksheet interface for mixing input, output and annotation in a document ready for publication, and therefore, can be used as:

a) automated mathematician (or mathematical assistant), for specifying the mathematical formulation in a manner close to the standard scientific notation

b) a numerical and symbolic calculator

c) a visualization system for functions and data (including two-dimensional and three-dimensional plots, and possibly, animation facilities), and

d) an environment to create interactive documents that mix text, formulas, animated graphics and sound.

Among the advantages of using symbolic computations in structural mechanics, the following can be cited:

a) significantly reducing the tedium of algebraic manipulations in the early stages of generation the mathematical model and the structural response,

b) providing the framework for describing the mathematical model in a suitable format for manipulation in subsequent phases, thereby allowing the analytical work to be pushed further before numerical computations start, thereby providing a new dimension for analytical and hybrid analytical/numerical techniques.

c) enhancing the understanding of the mathematical model, and value of qualitative analysis

d) allowing the generalization of standard techniques,

e) allowing the automatic (or semi-automatic) generation of the code used in the analysis of the structure, and

f) increasing the reliability of the analysis.

The number of reported applications of computerized symbolic manipulation in structural mechanics is less than that in other fields of science and engineering. This is to be contrasted with the extensive applications of numerical computations reported in structural mechanics. For the benefit of the readers of this book, all the monographs, textbooks and conference proceedings that are known to the author are listed in Appendix V.

The objective of this paper is to give a brief review of the status and some recent applications of computerized symbolic manipulation systems in structural mechanics. The applications considered are: deriving the governing nonlinear differential equations for shallow shells; derivation of the stationary conditions for variational functionals; application of Rayleigh-Ritz and Bubnov-Galerkin techniques; and generation of characteristic arrays for finite elements. Other reported applications of computerized symbolic manipulation in structural mechanics are discussed, and future directions for research are outlined.

2. SELECTED RECENT APPLICATIONS

Reviews of some of the structural mechanics applications of computerized symbolic manipulation up to the late 1970s are given in Refs. [14], [15] and [16]. The most significant change since then is the wide availability of a number of symbolic computation systems on the new powerful engineering workstations, personal computers, and even hand-held computers (see Tables 1 and 2). In this section, application of computerized symbolic manipulation to a number of selected structural mechanics problems is described. In all these applications, the symbolic manipulation system Macsyma is used, and a listing of the Macsyma program is given. Very simple commands are used in these programs to demonstrate the ease of using symbolic computation systems.

2.1 Generation of the Governing Nonlinear Differential Equations for Shallow Shells

The governing differential equations for the displacement formulation of any structural system are obtained by combining three sets of fundamental equations: a) strain-displacement relations; b) constitutive relations; and c) equations of motion (or equilibrium) (see Fig. 1). The interactive Macsyma program for generating the governing nonlinear differential equations for shallow shells is given in Appendix I. Note that the constitutive relations are linear, but because of the assumption of large displacements, both the strain-displacement relations and the equilibrium equations are nonlinear.

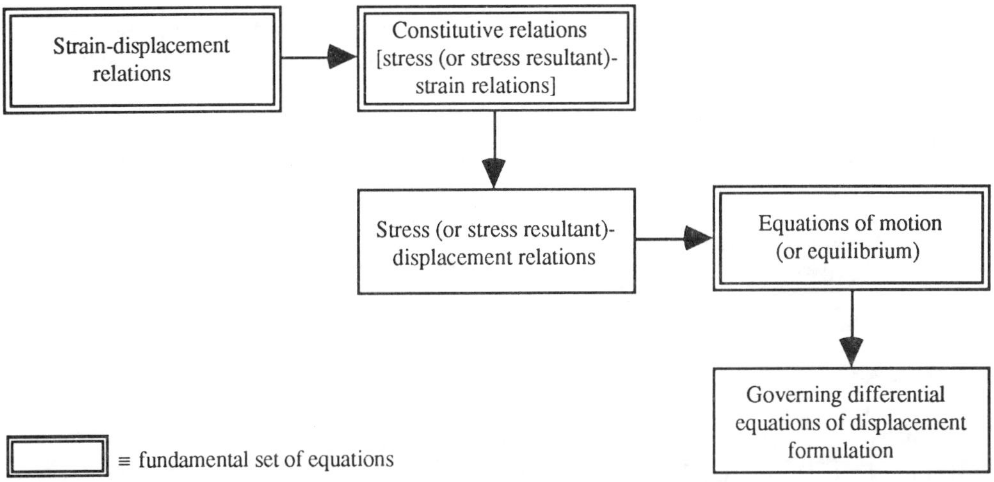

Figure 1 - Schematic representation of the steps used in generating the governing differential equations of the displacement formulation.

2.2 Derivation of the Stationary Conditions of Variational Functionals

Functionals are functions of other functions. The role of independent variables is played by functions. Examples are potential and kinetic energies. The stationary conditions of the functional consist of: a) Euler-Lagrange equation(s) in the domain; and b) the geometric (or kinematic) and natural boundary conditions at the boundary. If the functional fully characterizes the structural response, then the stationary condition(s) provide: a) the governing differential equation(s) that must be satisfied inside the domain of the structural member, and b) all possible conditions at the boundary. Two examples are presented herein. The first is the transverse bending of a beam on elastic foundation, and the second is the transverse bending of a shear flexible anisotropic plate. The functionals for both problems, along with the stationary conditions, are given subsequently. The interactive Macsyma programs are given in Appendix II (Parts a and b).

2.2.1 Transversely Loaded Beam on Elastic Foundation

If a displacement formulation is used, with the fundamental unknown selected to be the transverse displacement component w, the functional characterizing the response is the total potential energy p(w), given by:

$$\pi(w) = \int_{0}^{L} F\left(x,\ w,\ \frac{d^2w}{dx^2}\right) dx \tag{2.1}$$

$$F = \frac{1}{2}\, EI \left(\frac{d^2w}{dx^2}\right)^2 + \frac{1}{2}\, \kappa\, w^2 - p\, w \tag{2.2}$$

where EI and k are the bending stiffness and foundation modulus, and are both assumed to be functions of x; and p is the intensity of the transverse loading. The stationary conditions for the functional given by Eq. 2.1 are:

<u>Euler-Lagrange Equation</u>

$$\frac{\partial F}{\partial w} - \frac{d}{dx}\left[\frac{\partial F}{\partial\left(\frac{dw}{dx}\right)}\right] + \frac{d^2}{dx^2}\left[\frac{\partial F}{\partial\left(\frac{d^2w}{dx^2}\right)}\right] = 0 \tag{2.3}$$

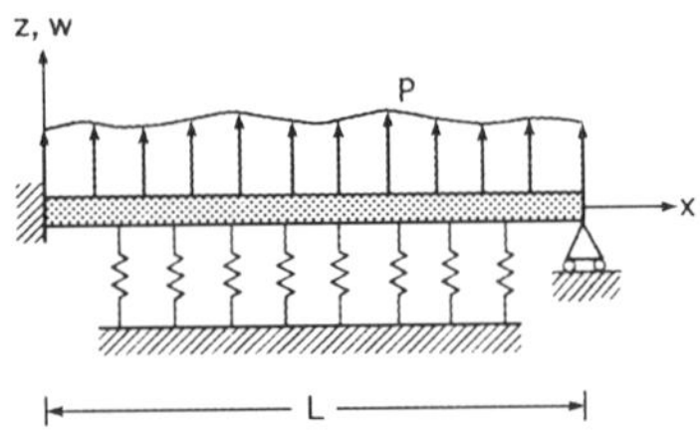

<u>Boundary Conditions</u>

At x=0 and x=L either

$$\left[\left(\frac{\partial F}{\partial\left(\frac{dw}{dx}\right)}\right) - \frac{d}{dx}\left(\frac{\partial F}{\partial\left(\frac{d^2w}{dx^2}\right)}\right)\right] = 0 \qquad \text{or } w \text{ is prescribed} \tag{2.4}$$

and either

$$\frac{\partial F}{\partial\left(\frac{d^2w}{dx^2}\right)} = 0 \qquad \text{or } \frac{dw}{dx} \text{ is prescribed.} \tag{2.5}$$

2.2.2 Transversely Loaded Shear-Flexible Plate

A mixed formulation is used in conjunction with a first-order shear deformation theory. The fundamental unknowns are selected to be the five stress resultants $M_{\alpha\beta}$, Q_α ($\alpha, \beta = 1, 2$); and the three generalized displacements w, ϕ_α. The functional characterizing the response and the stationary conditions are given by:

Variational Functional

$$\pi\left(M_{\alpha\beta}, Q_\alpha, w, \phi_\alpha\right) = \int_S F\left(x_\alpha, M_{\alpha\beta}, Q_\alpha, w, \partial_\alpha w, \phi_\beta, \partial_\alpha \phi_\beta\right) dS \qquad (2.6)$$

where

$$F = M_{\alpha\beta}\, \partial_\alpha \phi_\beta + Q_\alpha \left(\phi_\alpha + \partial_\alpha w\right) - \frac{1}{2}\, G_{\alpha\beta\gamma\rho}\, M_{\alpha\beta}\, M_{\gamma\rho} - \frac{1}{2}\, F_{\alpha3\beta3}\, Q_\alpha\, Q_\beta - p\, w \qquad (2.7)$$

Euler-Lagrange Equations

$$\frac{\partial F}{\partial Q_\alpha} = 0 \qquad (2.8)$$

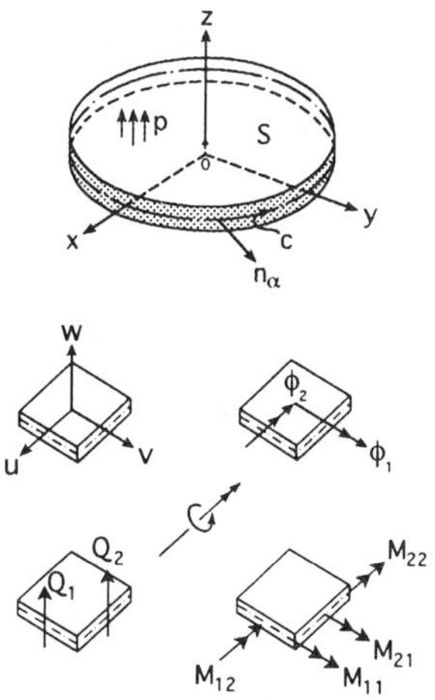

$$\frac{\partial F}{\partial M_{\alpha\beta}} = 0 \qquad (2.9)$$

$$\frac{\partial F}{\partial w} - \partial_\alpha \left(\frac{\partial F}{\partial\left(\partial_\alpha w\right)} \right) = 0 \qquad (2.10)$$

$$\frac{\partial F}{\partial \phi_\alpha} - \partial_\beta \left(\frac{\partial F}{\partial\left(\partial_\beta \phi_\alpha\right)} \right) = 0 \qquad (2.11)$$

Boundary Conditions

At each point of the boundary, either

$$\frac{\partial F}{\partial\left(\partial_\alpha w\right)}\, n_\alpha = 0 \qquad \text{or } w \text{ is prescribed} \qquad (2.12)$$

and either

$$\frac{\partial F}{\partial\left(\partial_\beta \phi_\alpha\right)}\, n_\beta = 0 \qquad \text{or } \phi_\alpha \text{ is prescribed.} \qquad (2.13)$$

In Eqs. 2.6-2.11, $G_{\alpha\beta\gamma\rho}$ and $F_{\alpha3\beta3}$ are the bending and transverse shear compliance coefficients of the plate, which are assumed to be independent of the coordinates x_1 and x_2; p is the intensity of the external transverse loading; and n_α is the unit outward normal to the boundary; and $\partial_\alpha \equiv \frac{\partial}{\partial x_\alpha}$.

2.3 Application of Rayleigh-Ritz Technique

Rayleigh-Ritz technique is a direct variational technique which can be used, in conjunction with any of the variational principles, to obtain approximate analytical solutions. The application of the Rayleigh-Ritz technique consists of: a) approximating each of the fundamental unknowns in the variational functional by a linear combination of preselected functions (trial functions or coordinate functions), each multiplied by an undetermined coefficient (amplitude of the coordinate function), and b) applying the stationary condition(s) of the discretized functional to obtain the undetermined coefficients (which provide the best combination of the preselected coordinate functions). Two examples are presented herein, namely: transversely loaded cantilever beam on elastic foundation, and transversely loaded plate with clamped edges. In the first example, the bending stiffness and foundation modulus are assumed to be constants, and the transverse displacement is approximated by a power series in x. In the second example, the Kirchhoff plate model is used and the transverse displacement is approximated by products of cosine polynomials. The approximations used for the transverse displacements in the two cases are given subsequently, and the Macsyma programs for evaluating the undetermined parameters are given in Appendix III (Parts a and b).

Displacement approximation for the cantilever beam on elastic foundation:

$$w = \sum_{i=1}^{4} A_i x^{i+1} \tag{2.14}$$

Displacement approximation for the square plate with clamped edges:

$$w = A_1 f_1(x) f_1(y) + A_2 f_1(x) f_2(y) + A_3 f_2(x) f_1(y) + A_4 f_2(x) f_2(y) \tag{2.15}$$

where

$$f_1(\ ) = 1 - \cos \frac{2\pi(\ )}{L} \tag{2.16}$$

$$f_2(\ ) = 1 - \cos \frac{2\pi(\ )}{L} - \cos \frac{4\pi(\ )}{L} \tag{2.17}$$

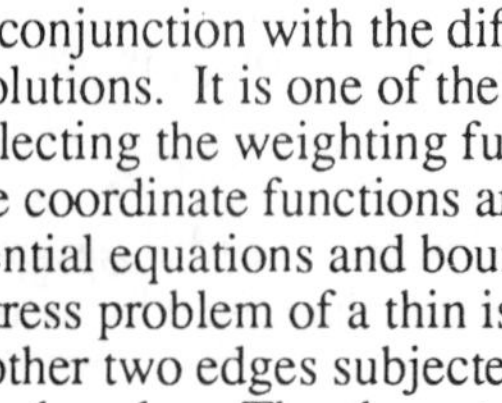

2.4 Application of Bubnov-Galerkin Technique

The Bubnov-Galerkin technique is typically used in conjunction with the differential equation formulation to obtain approximate analytical solutions. It is one of the family of weighted-residual techniques, and is characterized by selecting the weighting functions to be the trial (or coordinate) functions. The amplitudes of the coordinate functions are obtained by forming weighted averages of the residuals of differential equations and boundary conditions. The technique is applied herein to a plane stress problem of a thin isotropic square plate with two opposite edges clamped, and the other two edges subjected to prescribed normal traction with parabolic variation along the edge. The characteristics of the problem are given in Ref. [17] and are described subsequently. The interactive Macsyma program used for evaluating the amplitudes of the coordinate functions is given in Appendix IV.

<u>Boundary Conditions</u>

At $y = -1, +1$

$$u = v = 0 \tag{2.18}$$

At $x = -1, +1$

$$t_1 = E(1-y^2)/(1+v) \tag{2.19}$$

$$t_2 = 0 \tag{2.20}$$

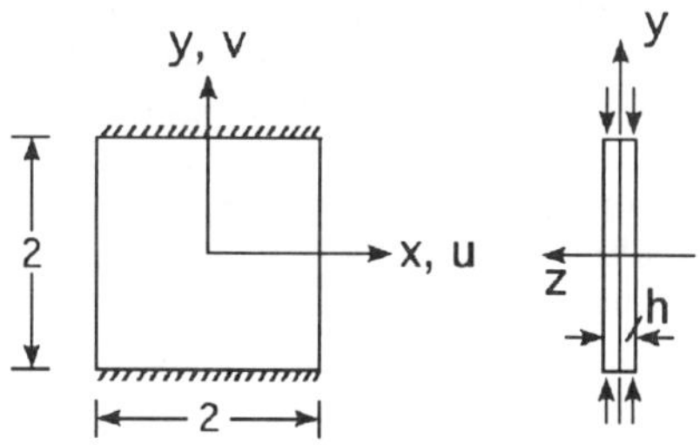

<u>Displacement Approximations</u>

$$u = x\left(1-y^2\right)\left[1 \quad x^2 \quad y^2\right]\begin{Bmatrix} A_1 \\ A_2 \\ A_3 \end{Bmatrix} \tag{2.21}$$

$$v = y\left(1-y^2\right)\left[1 \quad x^2 \quad y^2\right]\begin{Bmatrix} A_4 \\ A_5 \\ A_6 \end{Bmatrix} \tag{2.22}$$

3. OTHER REPORTED APPLICATIONS

Since the 1970s, computerized symbolic manipulation has been used in the generation of the characteristic arrays for simple linear finite elements (see, for example, Refs. [18] to [20]). Later, its use expanded to more general, complex linear and nonlinear finite elements (Refs. [21] to [33]), as well as to a wide range of other structural mechanics problems. Some of the early applications of computerized symbolic manipulation in structural mechanics are reviewed in Refs. [14], [15] and [16]. Variational and simple finite element applications are described in a monograph (Ref. [34]). Other reported applications include: 1) evaluation of effective stiffness and mass coefficients of continuum models for repetitive lattice structures (Ref. [16]); 2) interaction between axisymmetric and nonsymmetric creep buckling of circular cylinders (Ref. [35]); 3) application of Rayleigh-Ritz technique to static stress analysis, free vibraton; local and overall buckling of thin-walled structures (Refs. [36] to [38]); 4) application of a hybrid perturbation/Bubnov-Galerkin technique to the geometrically nonlinear analysis of curved beams (Ref. [39]); 5) constitutive modeling of structural materials (Refs. [40] to [44]); 6) determination of buckling loads and investigating dynamic stability for nonconservative systems (Refs. [45] to [48]); 7) random vibrations of structures (Ref. [49]); and 8) sensitivity analysis (Ref. [39]).

The symbolic computation facilitues used in these applications include: elementary matrix/vector operations; partial differentiation; factoring and other means of simplifications of expressions; solution of linear algebraic equations (in symbolic form); Taylor series expansion; integration by pattern matching; generation of FORTRAN (and C) codes; graphical displays and other advanced visualization facilities.

4. A LOOK AT THE FUTURE

Advances in computer science, computer technology and mathematics are likely to lead to new computing environment with the following characteristics:

• scientific workstations with well-integrated symbolic, wordprocessing, numeric and visualization capabilities, and multimedia facilities

• facilities for symbolic computation systems to interact with one another

• modular, reusable, library-style symbolic computation software

• efficient symbolic computation software for parallel architectures

• advanced user-interface facilities with a variety of software including text processors, numeric languages, object-oriented programming languages and AI programs.

• software tools to facilitate rapid, high-level creation of engineering system simulations.

The future computing environment, in addition to increasing the productivity of the engineer, will allow much higher levels of sophistication in the modeling, analysis and design of structures, as well as innovative applications of computerized symbolic manipulation to structural mechanics problems. These, in turn, can significantly contribute to the development of computational structures technology.

REFERENCES

1. Kahrimanian, H.G.: Analytic Differentiation by a Digital Computer, M.A. Thesis, Temple University, Philadelphia, PA, May 1953.
2. Nolan, J.: Analytic Differentiation on a Digital Computer, M.A. Thesis, Mathematics Dept., Massachusetts Institute of Technology, Cambridge, MA, May 1953.
3. Boyle, A. and Caviness, B.F. (Eds.): Future Directions for Research in Symbolic Computation, Report of a Workshop on Symbolic and Algebraic Computation, April 29-30, 1988, Washington, D.C., Society for Industrial and Applied Mathematics, Philadelphia, PA, 1990.
4. Bau, H.H., Herbert, T. and Yovanovich, M.M. (Eds.): Symbolic Computation in Fluid Mechanics and Heat Transfer, HTD Vol. 105, AMD Vol. 97, The American Society of Mechanical Engineers, New York, 1988.
5. Noor, A.K., Elishakoff, I. and Hulbert, G. (Eds.): Symbolic Computations and Their Impact on Mechanics, PVP Vol. 205, The American Society of Mechanical Engineers, New York, 1990.
6. Pavelle, R. and Wang, P.S.: MACSYMA from F to G, Symbolic Computation, 1 (1985), 69-100.
7. Foster, K.R. and Bau, H.H.: Symbolic manipulation programs for the personal computer, Science, 243 (1989).
8. Fateman, R.J.: A review of Macsyma, IEEE Trans. on Knowledge & Data Eng., 1(1) (1989), 133-145.
9. Simon, B.: Symbolic math software - It's not just for mainframes anymore, PC Magazine (Aug. 1992).
10. Simon, B.: Comparative CAS reviews, Notices of the American Mathematical Society, 39(7) (1992), 696-706.
11. Abbott, P.: Maple V and Mathematica, Notices of the American Mathematical Society, 39(8) (1992), 829-837.
12. Fateman, R.J.: A review of Mathematica, J. Symbolic Computation, 13 (1992), 545-579.

13. Foster, K.R.: Macsyma for windows, IEEE Spectrum, 30(4) (1993), 66.
14. Levi, I.M.: Symbolic algebra by computer - applications to structural mechanics, 12th AIAA/ASME Structures, Structural Dynamics and Materials Conference, Anaheim, CA, April 19-21, 1971, AIAA Paper 71-363.
15. Jensen, J. and Niordson, F.: Symbolic and Algebraic Manipulation Languages and Their Application in Mechanics, in: Structural Mechanics Software Series, Vol. 1, University Press of Virginia, Charlottesville, VA, 1977, 541-576.
16. Noor, A.K. and Andersen, C.M.: Computerized symbolic manipulation in structural mechanics - progress and potential, Computers & Structures, 10 (1979), 95-118.
17. Zienkiewicz, O.C. and Morgan, K.: Finite Elements and Approximation, John Wiley & Sons, New York, 1983, 85-88.
18. Chepurniy, N.: Evaluation of high-order polynomial triangular finite elements using FORMAC, Proc. Second Symposium on Symbolic Algebraic Manipulation, 1971, 365-371.
19. Pedersen, P. and Megahed, M.M.: Axisymmetric element analysis using analytical computing, Computers & Structures, 5 (1975), 241-247.
20. Pedersen, P.: On computer-aided analytic element analysis and the similarities of tetrahedron elements, Int. J. Numer. Meth. Eng., 11 (1977), 611-622.
21. Andersen, C.M. and Noor, A.K.: Use of Group-Theoretic Methods in the Development of Nonlinear Shell Finite Elements, Proc. Int. Conference on Symmetry, Similarity and Group-Theoretic Meth. Mech., University of Calgary, Calgary, Canada, 1974, 533-558.
22. Andersen, C.M. and Noor, A.K.: A Computerized Symbolic Integration Technique for Development of Triangular and Quadrilateral Composite Shallow-Shell Finite Elements, NASA TN-D-8067, 1975.
23. Noor, A. K. and Andersen, C. M.: Mixed isoparametric finite element models of laminated composite shells, Comp. Meth. Appl. Mech. Eng., 11 (1977), 255-280.
24. Korncoff, A.R. and Fenves, S.J.: Symbolic generation of finite element stiffness matrices, Computers & Structures, 10 (1979), 119-124
25. Noor, A.K. and Andersen, C.M.: Computerized symbolic manipulation in nonlinear finite element analysis, Computers & Structures, 13 (1981), 379-403.
26. Wang, P.W.: FINGER: A symbolic system for automatic generation of numerical programs in finite element analysis, J. Symbolic Computation, 2 (1986), 305-316.
27. Wang, P.W., Tan, H., Saleeb, A. and Chang, T.Y.: Code generation for hybrid mixed mode formulation in finite element analysis, Proc. ACM SYMSAC '88 Conf., Univ. of Waterloo, Canada (1986), 45-51.
28. Barbier, C., Clark, P.J., Bettess, P. and Bettess, J.A.: Automatic generation of shape functions for finite element analysis using REDUCE, Eng. Comput., 7(4) (1990), 349-358.
29. Chang, T.Y.P. and Tan, H.Q.: Application of Symbolic Computations in Finite Element Structural Analysis, in: Symbolic Computations and Their Impact on Mechanics (Eds. A.K. Noor, I. Elishakoff, G. Hulbert), PVP Vol. 205, Proc. symposium, ASME Winter Annual Meeting, Nov. 25-30, 1990, Dallas, TX, American Society of Mechanical Engineers, New York, 1990, 205-213.
30. Chang, T.Y., Tan, H.Q., Zheng, D. and Yuan, M.W.: Application of symbolic method to hybrid/mixed finite elements and computer implementation, Computers & Structures, 35(4) (1990), 293-299.

31. Tan, H.Q., Chang, T.Y.P. and Zheng, D.: On symbolic manipulation and code generation of a hybrid three-dimensional solid element, Engineering with Computers, 7(1) (1991), 47-59.
32. Nagabhushanam, J., Srinivas, C.J. and Gaonkar, G.H.: Symbolic generation of elemental matrices for finite Element Analysis, Computers & Structures, 42(3) (1992), 375-380.
33. Sharma, N. and Wang, P.S.: The PIER parallel FEA program generator, Proc. Thirteenth Annual ASME Int. Conf. in Engineering Conf. and Exposition, San Diego, CA (1993), 295-306.
34. Beltzer, A.I.: Variational and Finite Element Methods - A Symbolic Computation Approach, Springer-Verlag, Berlin, 1990.
35. Levi, I.M. and Hoff, N. J.: Interaction Between Axisymmetric and Nonsymmetric Creep Buckling of Circular Cylindrical Shells in Axial Compression, SUDAAR Rept. No. 388, Dept. of Aeronautics and Astronautics, Stanford University, Stanford, CA, 1970.
36. Andersen, C.M. and Noor, A.K.: Free Vibration of Laminated Composite Elliptic Plates, in: Advances in Engineering Science, Vol. 2, NASA CP-2001, 1976, 425-438.
37. Andersen, C. M. and Noor, A.K.: Symbolic Manipulation Techniques for Vibration Analysis of Laminated Elliptic Plates, in: Proc. 1977 MACSYMA Users' Conference, NASA CP-2012, 1977, 161-176.
38. Cohen, J.: Symbolic and numerical computer analysis of the combined local and overall buckling of rectangular thin-walled columns, Comp. Meth. Appl. Mech. Eng., 7 (1976), 17-38.
39. Noor, A.K. and Andersen, C.M.: Application of Symbolic Computation to Geometrically Nonlinear Analysis of Curved Beams," in: Symbolic Computations and Their Impact on Mechanics (Ed. A.K. Noor, I. Elishakoff and G. Hulbert), PVP Vol. 205, American Society of Mechanical Engineers, New York, 1990, 115-148.
40. Tan, H.Q.: Symbolic Derivation of Material Property Matrices in Finite Element Analysis, in: Symbolic Computation in Fluid Mechanics and Heat Transfer (Eds. H.H. Bau, T. Herbert and M.M. Yovanovich), HTD Vol. 105, American Society of Mechanical Engineers, New York, 1988, 111-116.
41. Arnold, S.M., Tan, H.Q. and Dong, X.: Computer Simulation of the Mathematical Modeling Involved in Constitutive Equation Development: Via Symbolic Computations, NASA TM-102532, 1989.
42. Arnold, S.M., Tan, H.Q. and Dong, X.: Application of Symbolic Computations to the Constitutive Modeling of Structural Materials, NASA TM-103225, 1990.
43. Arnold, S.M., Saleeb, A.F., Tan, H.Q. and Zhang, Y.: Explicit Robust Schemes for Implementation of a Class of Principal Value-Based Constitutive Models: Symbolic and Numerical Implementation, NASA TM-106124, 1993.
44. Arnold, S.M. and Tan, H.Q.: Symbolic derivation of potential based constitutive equations, Comput. Mech., 6(3) (1990), 237-246.
45. Elishakoff, I. and Couch, B.: Nonuniform Leipholz's column on elastic foundation - Instability study by symbolic algebra, SM Archives, 12(1) (1987), 379-389.
46. Elishakoff, I. and Hollkamp, J.: Computerized symbolic solution for a nonconservative system in which instability occurs by flutter in one range of a parameter and by divergence in another, Comp. Meth. Appl. Mech. Eng., 62 (1987), 27-46.
47. Elishakoff, I. and Wang, X.: Generalization of Smith-Herrmann problem with the aid of computerized symbolic algebra, J. Sound & Vibration, 117(3) (1987), 537-542.

48. Lottati, I. and Elishakoff, I.: Influence of Shear Deformation and Rotary Inertia on Flutter of a Cantilevered Beam-Exact and Symbolic Computerized Solutions, in: Refined Dynamical Theories in Beams, Plates and Shells and Their Applications (Eds. I. Elishakoff and H. Irretier), Springer-Verlag, Berlin, 1987, 261-273.
49. Rehak, M.L., DiMaggio, F.L., Benaroya, H. and Elishakoff, I.: Random vibrations with Macsyma, Comp. Meth. Appl. Mech. Eng., 61 (1987), 61-70.

Table 1 - List of Some of the Current General-Purpose Symbolic Manipulation Systems

System	Computers	Contact
Macsyma	Sun-3 & 4 (SPARC), VAX (UNIX, ULTRIX, VMS), Apollo, HP 9000, DEC RISC, Symbolics Computers,RS6000, PC 386/387 and 486	Macsyma, Inc. Arlington, VA
Mathematica	Macintosh (system 6.07 or later and A/UX) MS-DOS 386 and higher, Microsoft Windows, Unix workstations(Sun-4, HP, DEC, IBM Risc, Sony, Silicon Graphics, 386-Unix), NeXT, CONVEX and MIPS	Wolfram Res., Inc. Champaign, IL
Maple V	Unix workstations (DEC, HP, IBM, MIPS, Sun, SGI, Apollo), PC-386 (DOS and Unix), Mac, Amiga, Atari, AT&T 3B2, Gould, Convex, NCR, Pyramid, Sequent, Unisys and Cray's	Waterloo Maple Software Waterloo, Ont., Canada
Reduce	Cray Y-MP down to Mac's, PC's and Atari's	RAND Corp. Santa Monica, CA
Axiom	Sun Sparc, IBM RS 6000 and other IBM mainfram platforms	Numerical Algorithms Group, Inc. Downers Grove, IL
Derive	PC's and HP 95	Soft Warehouse, Inc. Honolulu, HI
FLAC	IMB PC's	Inst. Control Sciences Moscow, Russia
FORM	MS-DOS, Atarist, Amiga, Mac, Sun 3, Sun Sparc, Apollo, NeXT, DEC Station, HP, IBM Risc, IBM 3090, MIPS, Alliant, Gould, SGI	CAN, Amsterdam The Netherlands
MuPAD	Unix workstations (IBM, Sun), PC-386 (DOS and Unix), Mac, Sequent	MuPAD Distribution Univ. of Paderborn, Germany
SENAC	Sun 3, Sun 4 (SPARC), VAX (UNIX, ULTRIX VMS), PC	Univ. of Waikato Hamilton, New Zealand
Theorist	Mac's	Prescience Corporation San Francisco, CA

Table 2 - List of Some of the Specialized Symbolic Manipulaton Systems

System	Brief Description	Computers	Contact
		a) Tensor Analysis	
MathTensor	Elementary and advanced tensor analysis. Adds more than 250 functions and objects to Mathematica	All computers that run Mathematica 1.2, 2.0 or higher	MathSolutions, Inc. Chapel Hill, NC
Ricci	Mathematica package for doing tensor computations in differential geometry	All computers that run Mathematica 2.0 or higher	Dept. of Mathematics University of Washington Seattle, WA
STENSOR	System for tensor calculus and non-commutative algebra. Includes the systems SHEEP/CLASSI	IBM 386, 486, Sun, Apollo, Orion, Atari, Amiga, VAX DECStation, IBM 3090	Physics Dept., University of Stockholm, Sweden
GRTensor		All computers that run MapleV	Dept. of Physics, Queen's Univ., Kingston, Ontario, Canada
		b) Differential Equations	
ODE	Differential equation computations with easy-to-use menus and mouse support	IBM PC's (DOS)	Dept. of Applied Math., University of Adelaide, Australia
DELiA	Differential equation computations	IBM PC's (DOS)	Wolfram Research Urbana-Champaign, IL

APPENDIX I

```
/*****************************************************************************
*                                                                           *
* MACSYMA PROGRAM FOR GENERATING THE GOVERNING DIFFERENTIAL                 *
*      EQUATIONS FOR LARGE DEFLECTION ANALYSIS OF LAMINATED                 *
*      ANISOTROPIC SHALLOW SHELLS-DISPLACEMENT FORMULATION                  *
*                                                                           *
*****************************************************************************/
```

/* *Fundamental unknowns - displacements* */

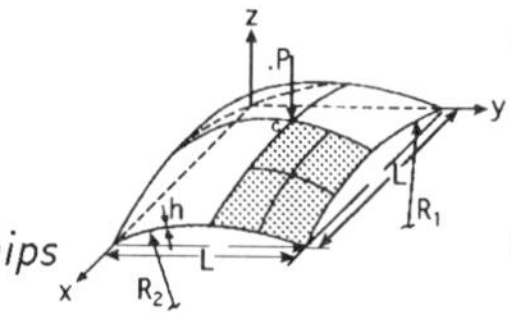

(C1) DEPENDS ([U,V,W], [X,Y]);

$$[u(x, y),\ v(x, y),\ w(x, y)]$$

/* *Shell Strains – strain - displacement relationships* */

(C2) E1 : DIFF(U,X) + K1*W + DIFF(W,X)^2/2 ;

$$\frac{1}{2}\left(\frac{dw}{dx}\right)^2 + \mathrm{k1}\,w + \frac{du}{dx}$$

(C3) E2 : DIFF(V,Y) + K2*W + DIFF(W,Y)^2/2 ;

$$\frac{1}{2}\left(\frac{dw}{dy}\right)^2 + \mathrm{k2}\,w + \frac{dv}{dy}$$

(C4) GM12 : DIFF(U,Y) + DIFF(V,X) + 2*K12*W + DIFF(W,X)*DIFF(W,Y);

$$\frac{dw}{dx}\frac{dw}{dy} + 2\,\mathrm{k12}\,w + \frac{dv}{dx} + \frac{du}{dy}$$

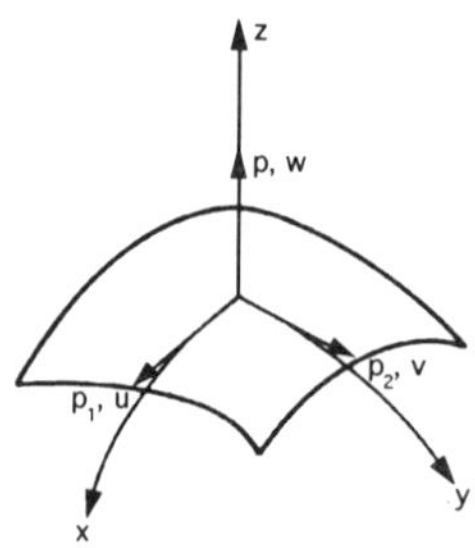

(C5) KAP1 : -DIFF(W,X,2);

$$-\frac{d^2 w}{dx^2}$$

(C6) KAP2 : -DIFF(W,Y,2);

$$-\frac{d^2 w}{dy^2}$$

(C7) TKAP12 : -2*DIFF(W,X,1,Y,1);

$$-2\,\frac{d^2 w}{dx\,dy}$$

/* *Shell Stress Resultants - Constitutive relations* */

```
(C8) CC: MATRIX(
  [ A11, A12, A16, B11 , B12 , B16  ],
  [ A12, A22, A26, B12 , B22 , B26  ],
  [ A16, A26, A66, B16 , B26 , B66  ],
  [ B11, B12, B16, DD11, DD12, DD16 ],
  [ B12, B22, B26, DD12, DD22, DD26 ],
  [ B16, B26, B66, DD16, DD26, DD66 ] )$
(C9) EE: [ E1, E2, GM12, KAP1, KAP2, TKAP12 ]$
(C10) NN: CC.TRANSPOSE(EE)$
(C11) N1   : NN[1]$
(C12) N2   : NN[2]$
(C13) N12  : NN[3]$
(C14) M1   : NN[4]$
(C15) M2   : NN[5]$
(C16) M12  : NN[6]$
```

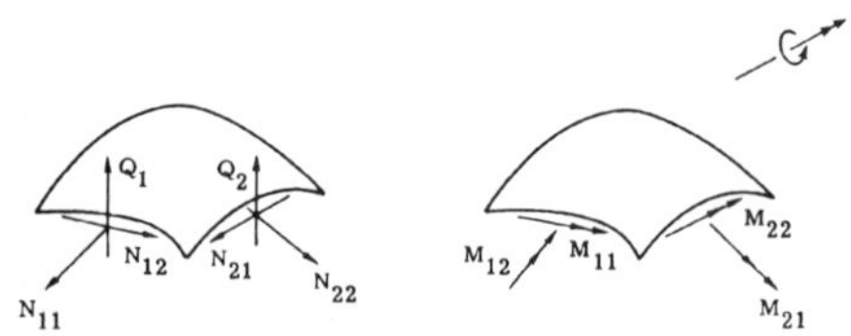

/* *Governing equations - Equilibrium equations* */

(C17) EQ[1] : DIFF(N1 ,X) + DIFF(N12,Y) - P1;

$$-\mathrm{b26}\,\frac{d^3 w}{dy^3} + \mathrm{a26}\left(\frac{dw}{dy}\frac{d^2 w}{dy^2} + \mathrm{k2}\,\frac{dw}{dy} + \frac{d^2 v}{dy^2}\right) + \mathrm{a66}\left(\frac{dw}{dx}\frac{d^2 w}{dy^2} + \frac{d^2 w}{dx\,dy}\frac{dw}{dy} + 2\,\mathrm{k12}\,\frac{dw}{dy} + \frac{d^2 v}{dx\,dy} + \frac{d^2 u}{dy^2}\right)$$

$$+\mathrm{a16}\left(\frac{d^2 w}{dx^2}\frac{dw}{dy} + \frac{dw}{dx}\frac{d^2 w}{dx\,dy} + 2\,\mathrm{k12}\,\frac{dw}{dx} + \frac{d^2 v}{dx^2} + \frac{d^2 u}{dx\,dy}\right) + \mathrm{a12}\left(\frac{d^2 w}{dx\,dy}\frac{dw}{dy} + \mathrm{k2}\,\frac{dw}{dx} + \frac{d^2 v}{dx\,dy}\right)$$

$$+\mathrm{a16}\left(\mathrm{k1}\,\frac{dw}{dy}+\frac{dw}{dx}\,\frac{d^2w}{dxdy}+\frac{d^2u}{dxdy}\right)-\mathrm{b11}\,\frac{d^3w}{dx^3}-3\,\mathrm{b16}\,\frac{d^3w}{dx^2dy}+\mathrm{a11}\left(\frac{dw}{dx}\,\frac{d^2w}{dx^2}+\mathrm{k1}\,\frac{dw}{dx}+\frac{d^2u}{dx^2}\right)$$
$$-2\,\mathrm{b66}\,\frac{d^3w}{dxdy^2}-\mathrm{b12}\,\frac{d^3w}{dxdy^2}-\mathrm{p1}$$

(C18) EQ[2] : DIFF(N12,X) + DIFF(N2 ,Y) - P2;

$$-\mathrm{b22}\,\frac{d^3w}{dy^3}+\mathrm{a22}\left(\frac{dw}{dy}\,\frac{d^2w}{dy^2}+\mathrm{k2}\,\frac{dw}{dy}+\frac{d^2v}{dy^2}\right)+\mathrm{a26}\left(\frac{dw}{dx}\,\frac{d^2w}{dy^2}+\frac{d^2w}{dxdy}\,\frac{dw}{dy}+2\,\mathrm{k12}\,\frac{dw}{dy}+\frac{d^2v}{dxdy}+\frac{d^2u}{dy^2}\right)$$
$$+\mathrm{a66}\left(\frac{d^2w}{dx^2}\,\frac{dw}{dy}+\frac{dw}{dx}\,\frac{d^2w}{dxdy}+2\,\mathrm{k12}\,\frac{dw}{dx}+\frac{d^2v}{dx^2}+\frac{d^2u}{dxdy}\right)+\mathrm{a26}\left(\frac{d^2w}{dxdy}\,\frac{dw}{dy}+\mathrm{k2}\,\frac{dw}{dx}+\frac{d^2v}{dxdy}\right)$$
$$+\mathrm{a12}\left(\mathrm{k1}\,\frac{dw}{dy}+\frac{dw}{dx}\,\frac{d^2w}{dxdy}+\frac{d^2u}{dxdy}\right)$$
$$-\mathrm{b16}\,\frac{d^3w}{dx^3}-2\,\mathrm{b66}\,\frac{d^3w}{dx^2dy}-\mathrm{b12}\,\frac{d^3w}{dx^2dy}+\mathrm{a16}\left(\frac{dw}{dx}\,\frac{d^2w}{dx^2}+\mathrm{k1}\,\frac{dw}{dx}+\frac{d^2u}{dx^2}\right)-3\,\mathrm{b26}\,\frac{d^3w}{dxdy^2}-\mathrm{p2}$$

(C19) EQ[3] :-K1*N1 - K2*N2 -2*K12*N12 + DIFF(M1 ,X,2) + DIFF(M2 ,Y,2) + 2*DIFF(M12,X,1,Y,1) - P3;

$$-\mathrm{dd22}\,\frac{d^4w}{dy^4}+\mathrm{b22}\left(\frac{dw}{dy}\,\frac{d^3w}{dy^3}+\left(\frac{d^2w}{dy^2}\right)^2+\mathrm{k2}\,\frac{d^2w}{dy^2}+\frac{d^3v}{dy^3}\right)$$
$$+\mathrm{b26}\left(\frac{dw}{dx}\,\frac{d^3w}{dy^3}+2\,\frac{d^2w}{dxdy}\,\frac{d^2w}{dy^2}+2\,\mathrm{k12}\,\frac{d^2w}{dy^2}+\frac{d^3w}{dxdy^2}\,\frac{dw}{dy}+\frac{d^3v}{dxdy^2}+\frac{d^3u}{dy^3}\right)$$
$$+2\,\mathrm{b66}\left(\frac{d^2w}{dx^2}\,\frac{d^2w}{dy^2}+\frac{d^3w}{dx^2dy}\,\frac{dw}{dy}+\frac{dw}{dx}\,\frac{d^3w}{dxdy^2}+\left(\frac{d^2w}{dxdy}\right)^2+2\,\mathrm{k12}\,\frac{d^2w}{dxdy}+\frac{d^3v}{dx^2dy}+\frac{d^3u}{dxdy^2}\right)$$
$$+2\,\mathrm{b26}\left(\frac{d^2w}{dxdy}\,\frac{d^2w}{dy^2}+\frac{d^3w}{dxdy^2}\,\frac{dw}{dy}+\mathrm{k2}\,\frac{d^2w}{dxdy}+\frac{d^3v}{dxdy^2}\right)-2\,\mathrm{dd16}\,\frac{d^4w}{dx^3dy}-4\,\mathrm{dd66}\,\frac{d^4w}{dx^2dy^2}$$
$$+2\,\mathrm{b16}\left(\frac{dw}{dx}\,\frac{d^3w}{dx^2dy}+\frac{d^2w}{dxdy}\,\frac{d^2w}{dx^2}+\mathrm{k1}\,\frac{d^2w}{dxdy}+\frac{d^3u}{dx^2dy}\right)-2\,\mathrm{dd26}\,\frac{d^4w}{dxdy^3}$$
$$+\mathrm{b12}\left(\mathrm{k1}\,\frac{d^2w}{dy^2}+\frac{dw}{dx}\,\frac{d^3w}{dxdy^2}+\left(\frac{d^2w}{dxdy}\right)^2+\frac{d^3u}{dxdy^2}\right)$$
$$-2\,\mathrm{k12}\left(-\mathrm{b26}\,\frac{d^2w}{dy^2}+\mathrm{a26}\left(.5\left(\frac{dw}{dy}\right)^2+\mathrm{k2}\,w+\frac{dv}{dy}\right)+\mathrm{a66}\left(\frac{dw}{dx}\,\frac{dw}{dy}+2\,\mathrm{k12}\,w+\frac{dv}{dx}+\frac{du}{dy}\right)\right)$$
$$-2\,\mathrm{k12}\left(-\mathrm{b16}\,\frac{d^2w}{dx^2}-2\,\mathrm{b66}\,\frac{d^2w}{dxdy}+\mathrm{a16}\left(.5\left(\frac{dw}{dx}\right)^2+\mathrm{k1}\,w+\frac{du}{dx}\right)\right)$$
$$-\mathrm{k2}\left(-\mathrm{b22}\,\frac{d^2w}{dy^2}+\mathrm{a22}\left(.5\left(\frac{dw}{dy}\right)^2+\mathrm{k2}\,w+\frac{dv}{dy}\right)+\mathrm{a26}\left(\frac{dw}{dx}\,\frac{dw}{dy}+2\,\mathrm{k12}\,w+\frac{dv}{dx}+\frac{du}{dy}\right)\right)$$
$$-\mathrm{k2}\left(-\mathrm{b12}\,\frac{d^2w}{dx^2}-2\,\mathrm{b26}\,\frac{d^2w}{dxdy}+\mathrm{a12}\left(.5\left(\frac{dw}{dx}\right)^2+\mathrm{k1}\,w+\frac{du}{dx}\right)\right)$$
$$-\mathrm{k1}\left(-\mathrm{b12}\,\frac{d^2w}{dy^2}+\mathrm{a12}\left(.5\left(\frac{dw}{dy}\right)^2+\mathrm{k2}\,w+\frac{dv}{dy}\right)+\mathrm{a16}\left(\frac{dw}{dx}\,\frac{dw}{dy}+2\,\mathrm{k12}\,w+\frac{dv}{dx}+\frac{du}{dy}\right)\right)$$
$$-\mathrm{k1}\left(-\mathrm{b11}\,\frac{d^2w}{dx^2}-2\,\mathrm{b16}\,\frac{d^2w}{dxdy}+\mathrm{a11}\left(.5\left(\frac{dw}{dx}\right)^2+\mathrm{k1}\,w+\frac{du}{dx}\right)\right)$$
$$+\mathrm{b16}\left(\frac{d^3w}{dx^3}\,\frac{dw}{dy}+\frac{dw}{dx}\,\frac{d^3w}{dx^2dy}+2\,\frac{d^2w}{dxdy}\,\frac{d^2w}{dx^2}+2\,\mathrm{k12}\,\frac{d^2w}{dx^2}+\frac{d^3v}{dx^3}+\frac{d^3u}{dx^2dy}\right)$$
$$+\mathrm{b12}\left(\frac{d^3w}{dx^2dy}\,\frac{dw}{dy}+\mathrm{k2}\,\frac{d^2w}{dx^2}+\left(\frac{d^2w}{dxdy}\right)^2+\frac{d^3v}{dx^2dy}\right)-\mathrm{dd11}\,\frac{d^4w}{dx^4}-2\,\mathrm{dd16}\,\frac{d^4w}{dx^3dy}$$
$$+\mathrm{b11}\left(\frac{dw}{dx}\,\frac{d^3w}{dx^3}+\left(\frac{d^2w}{dx^2}\right)^2+\mathrm{k1}\,\frac{d^2w}{dx^2}+\frac{d^3u}{dx^3}\right)-2\,\mathrm{dd12}\,\frac{d^4w}{dx^2dy^2}-2\,\mathrm{dd26}\,\frac{d^4w}{dxdy^3}-\mathrm{p3}$$

APPENDIX II

```
/*************************************************************************
*                                                                       *
*(a) MACSYMA PROGRAM FOR GENERATING THE GOVERNING DIFFERENTIAL *
*EQUATIONS AND NATURAL BOUNDARY CONDITIONS OF A TRANSVERSELY *
*     LOADED BEAM ON ELASTIC FOUNDATIONS – BOTH EI AND K ARE      *
*                           FUNCTIONS OF X                              *
*                                                                       *
*************************************************************************/
```

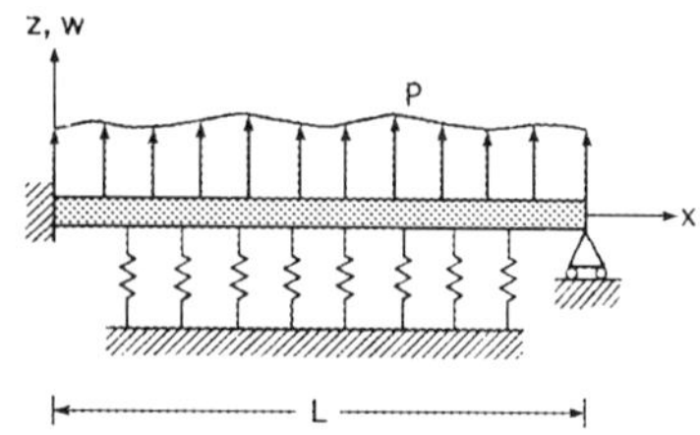


```
(C1) DEPENDS ( [W,EI,K], X );
```

$$[w(x), ei(x), k(x)]$$

```
(C2) FUNC::EI*DIFF(W,X,2)^2/2 + K*W^2/2 - P*W;
```

$$\frac{1}{2}\,ei\left(\frac{d^2w}{dx^2}\right)^2+\frac{k\,w^2}{2}-p\,w$$

```
(C3) EQ:DIFF(FUNC,W) - DIFF(DIFF(FUNC,DIFF(W,X)),X)
     + DIFF(DIFF(FUNC,DIFF(W,X,2)),X,2);
```

$$\mathrm{ei}\,\frac{d^4w}{dx^4}+2\,\frac{d\mathrm{ei}}{dx}\,\frac{d^3w}{dx^3}+\frac{d^2\mathrm{ei}}{dx^2}\,\frac{d^2w}{dx^2}+k\,w-p$$

```
(C4) BC1:DIFF(FUNC,DIFF(W,X)) - DIFF(DIFF(FUNC,DIFF(W,X,2)),X);
```

$$-\mathrm{ei}\,\frac{d^3w}{dx^3}-\frac{d\mathrm{ei}}{dx}\,\frac{d^2w}{dx^2}$$

```
(C5) BC2:DIFF(FUNC,DIFF(W,X,2));
```

$$\mathrm{ei}\,\frac{d^2w}{dx^2}$$

```
/*************************************************************************
*                                                                       *
*     (b) MACSYMA PROGRAM FOR GENERATING THE EULER-LAGRANGE        *
*EQUATIONS AND NATURAL BOUNDARY CONDITIONS FOR A TRANSVERSELY *
*        LOADED SHEAR FLEXIBLE PLATE – MIXED FORMULTATION           *
*                                                                       *
*************************************************************************/
```

```
(C1) DEPENDS([M11,M22,M12,Q1,Q2,W,PHI1,PHI2],[X,Y]);
```

$$[m11(x, y), m22(x, y), m12(x, y), q1(x, y), q2(x, y), w(x, y), phi1(x, y), phi2(x, y)]$$

```
/*          Curvature changes and transverse shear strains            */
/*          for the plate-strain-displacement relationships           */
```

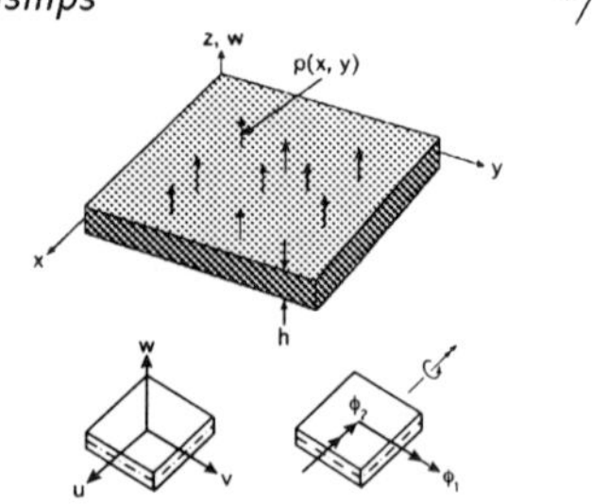


```
(C2) KAP1:DIFF(PHI1,X);
```

$$\frac{d\mathrm{phi1}}{dx}$$

```
(C3) KAP2:DIFF(PHI2,Y);
```

$$\frac{d\mathrm{phi2}}{dy}$$

```
(C4) TKAP12:DIFF(PHI1,Y)+DIFF(PHI2,X);
```

$$\frac{d\mathrm{phi2}}{dx}+\frac{d\mathrm{phi1}}{dy}$$

```
(C5) GAM13:PHI1+DIFF(W,X);
```

$$\frac{dw}{dx}+\mathrm{phi1}$$

```
(C6) GAM23:PHI2+DIFF(W,Y);
```

$$\frac{dw}{dy}+\mathrm{phi2}$$

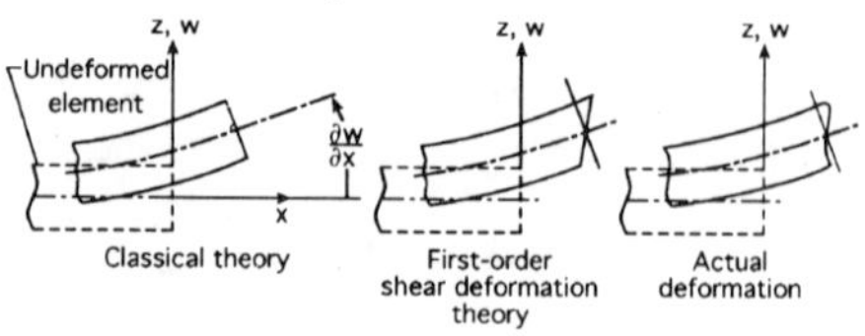


```
(C7) EE:[KAP1,KAP2,TKAP12,GAM13,GAM23];
```

$$\left\{\frac{dphi1}{dx}, \frac{dphi2}{dy}, \frac{dphi2}{dx} + \frac{dphi1}{dy}, \frac{dw}{dx} + phi1, \frac{dw}{dy} + phi2\right\}$$

/* *Stress resultants* */

(C8) MM:[M11,M22,M12,Q1,Q2];

[m11, m22, m12, q1, q2]

/* *Constitutive matrix for the plate* */

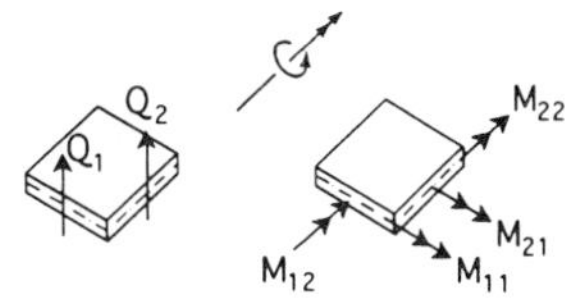


```
(C9) GG:MATRIX(
     [G11, G12, G16,  0,  0],
     [G12, G22, G26,  0,  0],
     [G16, G26, G66,  0,  0],
     [0,   0,   0,  F11, F12],
     [0,   0,   0,  F12, F22]);
```

$$\begin{pmatrix} g11 & g12 & g16 & 0 & 0 \\ g12 & g22 & g26 & 0 & 0 \\ g16 & g26 & g66 & 0 & 0 \\ 0 & 0 & 0 & f11 & f12 \\ 0 & 0 & 0 & f12 & f22 \end{pmatrix}$$

/* *Functional for Hellinger–Reissner Mixed Variational Principle* */

(C10) FUNC:MM.TRANSPOSE(EE) - MM.GG.TRANSPOSE(MM)/2 - P*W;

$$q2\left(\frac{dw}{dy} + phi2\right) + q1\left(\frac{dw}{dx} + phi1\right) - p\,w - q2\left(\frac{f22\,q2}{2} + \frac{f12\,q1}{2}\right) - q1\left(\frac{f12\,q2}{2} + \frac{f11\,q1}{2}\right)$$
$$+m22\,\frac{dphi2}{dy} + m12\left(\frac{dphi2}{dx} + \frac{dphi1}{dy}\right) + m11\,\frac{dphi1}{dx} - m12\left(\frac{g26\,m22}{2} + \frac{g66\,m12}{2} + \frac{g16\,m11}{2}\right)$$
$$-m22\left(\frac{g22\,m22}{2} + \frac{g26\,m12}{2} + \frac{g12\,m11}{2}\right) - m11\left(\frac{g12\,m22}{2} + \frac{g16\,m12}{2} + \frac{g11\,m11}{2}\right)$$

/* *Euler-Lagrange equations* */

```
(C11) FOR I THRU 5 DO (
      EQ[I]:DIFF(FUNC,MM[I]),
      DISPLAY  (EQ[I])  );
```

$$EQ_1 = \frac{dphi1}{dx} - g12\,m22 - g16\,m12 - g11\,m11$$
$$EQ_2 = \frac{dphi2}{dy} - g22\,m22 - g26\,m12 - g12\,m11$$
$$EQ_3 = \frac{dphi2}{dx} + \frac{dphi1}{dy} - g26\,m22 - g66\,m12 - g16\,m11$$
$$EQ_4 = \frac{dw}{dx} - f12\,q2 - f11\,q1 + phi1$$
$$EQ_5 = \frac{dw}{dy} - f22\,q2 - f12\,q1 + phi2$$

(C12) DIS:[W,PHI1,PHI2];

[w, phi1, phi2]

```
(C13) FOR I THRU 3 DO (
      EQ[I+5]:DIFF(FUNC,DIS[I])
         - DIFF(DIFF(FUNC,DIFF(DIS[I],X)),X)
         - DIFF(DIFF(FUNC,DIFF(DIS[I],Y)),Y),
      DISPLAY (EQ[I+5])  );
```

$$EQ_6 = -\frac{dq2}{dy} - \frac{dq1}{dx} - p$$
$$EQ_7 = q1 - \frac{dm12}{dy} - \frac{dm11}{dx}$$
$$EQ_8 = q2 - \frac{dm22}{dy} - \frac{dm12}{dx}$$

/* *Natural boundary conditions* */

```
(C14) FOR I THRU 3 DO (
      BC[I]:DIFF(FUNC,DIFF(DIS[I],X))*NX + DIFF(FUNC,DIFF(DIS[I],Y))*NY,
      DISPLAY(BC[I])    );
```

$$BC_1 = \mathrm{ny\,q2} + \mathrm{nx\,q1}$$
$$BC_2 = \mathrm{m12\,ny} + \mathrm{m11\,nx}$$
$$BC_3 = \mathrm{m22\,ny} + \mathrm{m12\,nx}$$

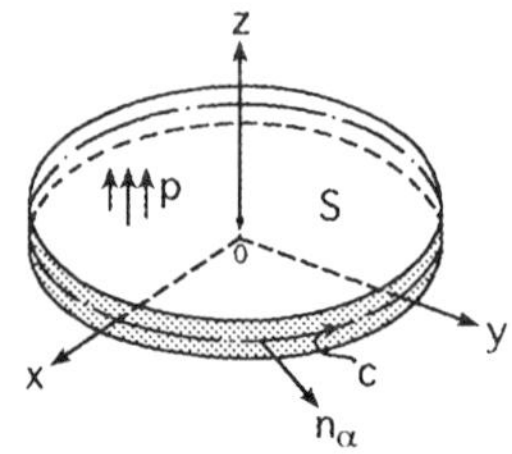

APPENDIX III

```
/*****************************************************************
*                                                                *
*      (a) MACSYMA PROGRAM FOR THE RAYLEIGH-RITZ ANALYSIS OF     *
*          TRANSVERSELY LOADED BEAM ON ELASTIC FOUNDATION        *
*                                                                *
*****************************************************************/
```

/* *Approximation for w* */

```
(C1) DEPENDENCIES(W(X));
```

[w(x)]

```
(C2) FF:[X^2, X^3, X^4, X^5];
```

$$[x^2, x^3, x^4, x^5]$$

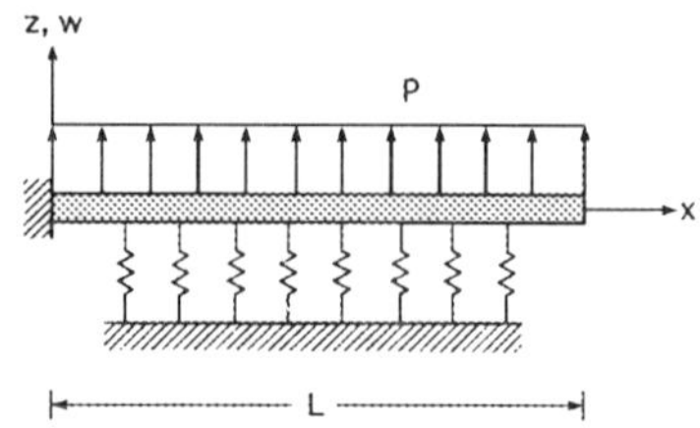


```
(C3) AA:[A1, A2, A3, A4];
```

[a1, a2, a3, a4]

```
(C4) W:AA.TRANSPOSE(FF);
```

$$\mathrm{a4}\,x^5 + \mathrm{a3}\,x^4 + \mathrm{a2}\,x^3 + \mathrm{a1}\,x^2$$

/* *Functional* */

```
(C5) FI:EI*DIFF(W,X,2)^2/2 + K*W^2/2 - P*W $
(C6) FUNC:INTEGRATE(FI,X,0,L)$
```

/* *Governing equations for A1, A2, ...* */

```
(C7) FOR I THRU 4 DO (
     EQ[I]: DIFF(FUNC,AA[I]),
     FOR J THRU I DO (
       K[I,J] : RATSIMP( DIFF(EQ[I],AA[J]) ),
       DISPLAY(K[I,J])
     )
   );
```

$$K_{1,1} = \frac{5544\,k\,l^5 + 110880\,\mathrm{ei}\,l}{27720}$$
$$K_{2,1} = \frac{4620\,k\,l^6 + 166320\,\mathrm{ei}\,l^2}{27720}$$
$$K_{2,2} = \frac{3960\,k\,l^7 + 332640\,\mathrm{ei}\,l^3}{27720}$$
$$K_{3,1} = \frac{3960\,k\,l^7 + 221760\,\mathrm{ei}\,l^3}{27720}$$
$$K_{3,2} = \frac{3465\,k\,l^8 + 498960\,\mathrm{ei}\,l^4}{27720}$$
$$K_{3,3} = \frac{3080\,k\,l^9 + 798336\,\mathrm{ei}\,l^5}{27720}$$
$$K_{4,1} = \frac{3080\,k\,l^9 + 798336\,\mathrm{ei}\,l^5}{27720}$$
$$K_{4,2} = \frac{3080\,k\,l^9 + 665280\,\mathrm{ei}\,l^5}{27720}$$
$$K_{4,3} = \frac{2772\,k\,l^{10} + 1108800\,\mathrm{ei}\,l^6}{27720}$$
$$K_{4,4} = \frac{2520\,k\,l^{11} + 1584000\,\mathrm{ei}\,l^7}{27720}$$

```
(C8) SUBLIST:[A1=0, A2=0, A3=0, A4=0];
                    [a1 = 0, a2 = 0, a3 = 0, a4 = 0]
(C9) FOR I THRU 4 DO (
       EQ[I]:DIFF(FUNC,AA[I]),
       P[I]:EXPAND(SUBST(SUBLIST,EQ[I])),
       P[I]: -P[I],
       DISPLAY(P[I])
   );
```

$$P_1 = \frac{l^3\,p}{3}$$
$$P_2 = \frac{l^4\,p}{4}$$
$$P_3 = \frac{l^5\,p}{5}$$
$$P_4 = \frac{l^6\,p}{6}$$

```
/*************************************************************
*                                                            *
*   (b) MACSYMA PROGRAM FOR THE RAYLEIGH-RITZ ANALYSIS OF    *
*       TRANSVERSELY LOADED SQUARE ISOTROPIC PLATE           *
*           CLASSICAL KIRCHHOFF PLATE THEORY                 *
*                                                            *
*************************************************************/
```

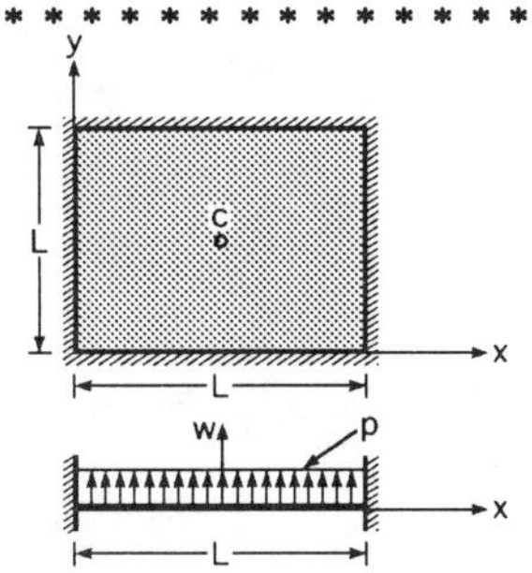


```
(C1) DEPENDENCIES(W(X,Y));
                    [w(x, y)]
/*                  Approximation for w                     */
(C2) PAI:%PI;
```

$$\pi$$

```
(C3) F1X:1-COS(2*PAI*X/L);
```

$$1 - \cos\left(\frac{2\,\pi\,x}{l}\right)$$

```
(C4) F1Y:1-COS(2*PAI*Y/L);
```

$$1 - \cos\left(\frac{2\,\pi\,y}{l}\right)$$

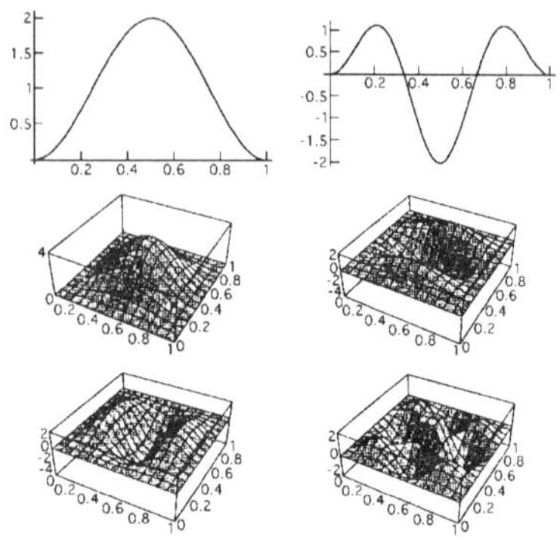

```
(C5) F2X:COS(2*PAI*X/L) - COS(4*PAI*X/L);
```

$$\cos\left(\frac{2\,\pi\,x}{l}\right) - \cos\left(\frac{4\,\pi\,x}{l}\right)$$

```
(C6) F2Y:COS(2*PAI*Y/L) - COS(4*PAI*Y/L);
```

$$\cos\left(\frac{2\,\pi\,y}{l}\right) - \cos\left(\frac{4\,\pi\,y}{l}\right)$$

```
(C7) FF:[F1X*F1Y, F1X*F2Y, F2X*F1Y, F2X*F2Y]$

(C8) AA:[A1,A2,A3,A4];
```

$$[a1, a2, a3, a4]$$

```
(C9) W:AA.TRANSPOSE(FF)$
```

/ Functional */*

```
(C10) FI1:D*(DIFF(W,X,2)+DIFF(W,Y,2))^2/2 $
(C11) FI2:-D*(1-NU)*(DIFF(W,X,2)*DIFF(W,Y,2) - DIFF(W,X,1,Y,1)^2) $
(C12) FI3:P*W$

(C13) FUNC1:INTEGRATE(FI1,X)$
(C14) FUNC1:SUBST(L,X,FUNC1) - SUBST(0,X,FUNC1)$
(C15) FUNC1:INTEGRATE(FUNC1,Y)$
(C16) FUNC1:RATSIMP(SUBST(L,Y,FUNC1) - SUBST(0,Y,FUNC1))$

(C17) FUNC2:INTEGRATE(FI2,X)$
(C18) FUNC2:SUBST(L,X,FUNC2) - SUBST(0,X,FUNC2)$
(C19) FUNC2:INTEGRATE(FUNC2,Y)$
(C20) FUNC2:RATSIMP( SUBST(L,Y,FUNC2) - SUBST(0,Y,FUNC2) );
```

$$0$$

```
(C21) FUNC3:INTEGRATE(FI3,X)$
(C22) FUNC3:SUBST(L,X,FUNC3) - SUBST(0,X,FUNC3)$
(C23) FUNC3:INTEGRATE(FUNC3,Y)$
(C24) FUNC3:RATSIMP( SUBST(L,Y,FUNC3) - SUBST(0,Y,FUNC3) )$
```

/ Governing equations for A1, A2, ... */*

```
(C25) FOR I THRU 4 DO (
        EQ[I]: DIFF(FUNC1,AA[I]),
        FOR J THRU 4 DO (
          K[I,J] : DIFF(EQ[I],AA[J]),
          DISPLAY(K[I,J])
        )
      );
```

$$K_{1,1} = \frac{32\,\pi^4\,d}{l^2}$$

$$K_{1,2} = -\frac{24\,\pi^4\,d}{l^2}$$

$$K_{1,3} = -\frac{24\,\pi^4\,d}{l^2}$$

$$K_{1,4} = \frac{16\,\pi^4\,d}{l^2}$$

$$K_{2,1} = -\frac{24\,\pi^4\,d}{l^2}$$

$$K_{2,2} = \frac{252\,\pi^4\,d}{l^2}$$

$$K_{2,3} = \frac{16\,\pi^4\,d}{l^2}$$

$$K_{2,4} = -\frac{116\,\pi^4\,d}{l^2}$$

$$K_{3,1} = -\frac{24\,\pi^4\,d}{l^2}$$

$$K_{3,2} = \frac{16\,\pi^4\,d}{l^2}$$

$$K_{3,3} = \frac{252\,\pi^4\,d}{l^2}$$

$$K_{3,4} = -\frac{116\,\pi^4\,d}{l^2}$$

$$K_{4,1} = \frac{16\,\pi^4\,d}{l^2}$$

$$K_{4,2} = -\frac{116\,\pi^4\,d}{l^2}$$

$$K_{4,3} = -\frac{116\,\pi^4\,d}{l^2}$$

$$K_{4,4} = \frac{472\,\pi^4\,d}{l^2}$$

```
(C26) SUBLIST:[A1=0, A2=0, A3=0, A4=0];
```

$$[a1 = 0,\ a2 = 0,\ a3 = 0,\ a4 = 0]$$

```
(C27) FOR I THRU 4 DO (
        P[I]:DIFF(FUNC3,AA[I]),
        DISPLAY(P[I])
    );
```

$$P_1 = l^2\,p$$

$$P_2 = 0$$

$$P_3 = 0$$

$$P_4 = 0$$

APPENDIX IV

```
/*************************************************************************
*                                                                       *
*   MACSYMA PROGRAM FOR THE BUBNOV GALERKIN ANALYSIS OF                 *
*   ISOTROPIC THIN PLATE SUBJECTED TO IN-PLANE EDGE TRACTIONS           *
*   PLANE STRESS PROBLEM - DISPLACEMENT FORMULATION                     *
*                                                                       *
*************************************************************************/
```

```
/*                  Approximations for u, v                             */
```

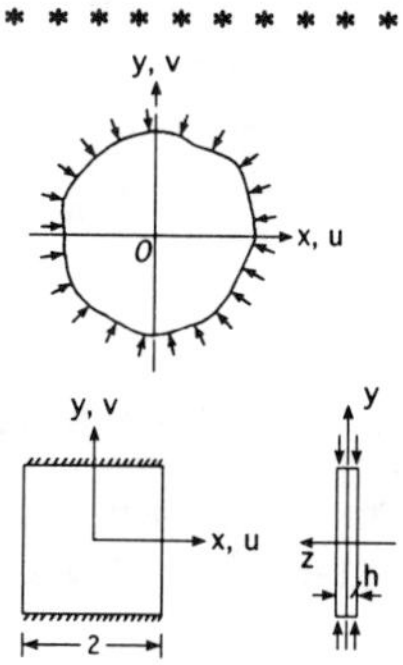


```
(C1) DEPENDS([U,V],[X,Y]) ;
```

$$[u(x,y)\,,v(x,y)]$$

```
(C2) FU1:X*(1-Y^2);
```

$$x\,(1-y^2)$$

```
(C3) FU2:X^3*(1-Y^2);
```

$$x^3\,(1-y^2)$$

```
(C4) FU3:X*Y^2*(1-Y^2);
```

$$x\,y^2\,(1-y^2)$$

(C5) FV1:Y*(1-Y^2);

$$y\ (1-y^2)$$

(C6) FV2:X^2*Y*(1-Y^2);

$$x^2\ y\ (1-y^2)$$

(C7) FV3:Y^3*(1-Y^2);

$$y^3\ (1-y^2)$$

(C8) FFU:[FU1,FU2,FU3];

$$[x\ (1-y^2)\ ,x^3\ (1-y^2)\ ,x\,y^2\ (1-y^2)]$$

(C9) FFV:[FV1,FV2,FV3];

$$[y\ (1-y^2)\ ,x^2\,y\ (1-y^2)\ ,y^3\ (1-y^2)]$$

(C10) AAU:[A1,A2,A3];

$$[\mathrm{A1}, \mathrm{A2}, \mathrm{A3}]$$

(C11) AAV:[A4,A5,A6];

$$[\mathrm{A4}, \mathrm{A5}, \mathrm{A6}]$$

(C12) U:FFU.TRANSPOSE(AAU);

$$\mathrm{a3}\,x\,y^2\ (1-y^2)+\mathrm{a2}\,x^3\ (1-y^2)+\mathrm{a1}\,x\ (1-y^2)$$

(C13) V:FFV.TRANSPOSE(AAV);

$$\mathrm{a6}\,y^3\ (1-y^2)+\mathrm{a5}\,x^2\,y\ (1-y^2)+\mathrm{a4}\,y\ (1-y^2)$$

```
/*                    Strain components                    */
(C14) E1:DIFF(U,X)$
(C15) E2:DIFF(V,Y)$
(C16) GM12:DIFF(U,Y) + DIFF(V,X)$
(C17) EE:[E1,E2,GM12]$
/*       Constitutive matrix and stress components         */
(C18) CC:MATRIX(
     [1, NU, 0],
     [NU, 1, 0],
     [0,  0, (1-NU)/2]
       ) * E / (1-NU^2)$
(C19) CC1:SUBST(1/4, NU, CC);
```

$$\begin{pmatrix} \frac{16\,E}{15} & \frac{4\,E}{15} & 0 \\ \frac{4\,E}{15} & \frac{16\,E}{15} & 0 \\ 0 & 0 & \frac{2\,E}{5} \end{pmatrix}$$

```
(C20) SIG:CC1.TRANSPOSE(EE)$
(C21) S11:SIG[1,1]$
(C22) S22:SIG[2,1]$
(C23) S12:SIG[3,1]$
/*   Governing differential equations / Equilibrium equations   */
(C24) EQ1:DIFF(S11,X) + DIFF(S12,Y)$
(C25) EQ2:DIFF(S12,X) + DIFF(S22,Y)$
/*          Boundary conditions at x=+1 and x=-1.          */
/*          S11A is the applied boundary traction          */
(C26) S11A  : E*(1-Y^2)/(1+NU) $
(C27) S11A  : SUBST(1/4, NU, S11A)$
(C28) S11B1 : SUBST( 1,X,S11) - S11A$
(C29) S11BM1: SUBST(-1,X,S11) - S11A$
```

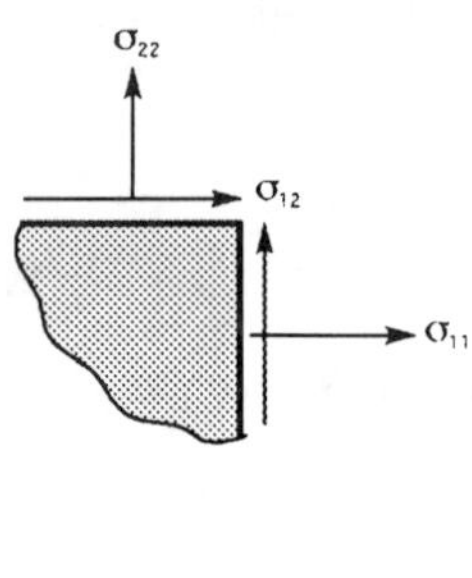

```
(C30) S12B1 : SUBST( 1,X,S12)$
(C31) S12BM1: SUBST(-1,X,S12)$
/*                    Bubnov Galerkin equations.                    */
/*           Weighted residuals of equilibrium equations            */
(C32) FOR I THRU 3 DO (
      EQ[I]: EQ1*FFU[I],
      EQ[I+3]: EQ2*FFV[I]
    );
(C33) FOR I THRU 6 DO (
      WR[I]: INTEGRATE(EQ[I],X,-1,1),
      WR[I]: INTEGRATE(WR[I],Y,-1,1)
    );
/*     Weighted residuals for boundary conditions at x=+1 and x=-1  */
(C34) FOR I THRU 3 DO (
      FFU1 : SUBST( 1, X, FFU[I]),
      FFUM1: SUBST(-1, X, FFU[I]),
      FFV1 : SUBST( 1, X, FFV[I]),
      FFVM1: SUBST(-1, X, FFV[I]),
      EQB[I  ] : -S11B1*FFU1 + S11BM1*FFUM1,
      EQB[I+3] : -S12B1*FFV1 + S12BM1*FFVM1,
      WRB[I  ] : INTEGRATE(EQB[I  ],Y,-1,1),
      WRB[I+3] : INTEGRATE(EQB[I+3],Y,-1,1)
    );
```

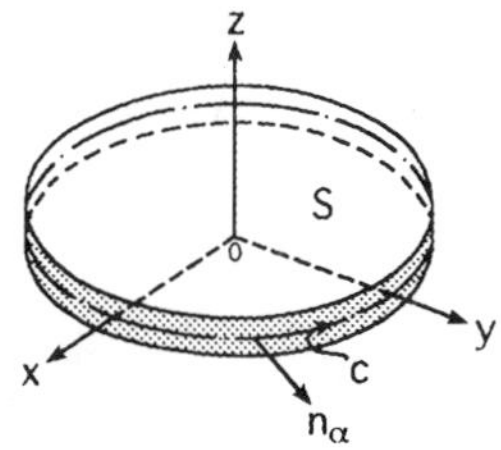


```
/*                    Equations for A1 ... A6                       */
/*                    Left hand side matrix                         */
(C35) AA:[A1,A2,A3,A4,A5,A6];
                         [A1, A2, A3, A4, A5, A6]
(C36) FOR I THRU 6 DO (
      FOR J THRU 6 DO (
        K[I,J]:DIFF(WR[I]+WRB[I], AA[J]),
      )
    );
(C37) SUBLIST:[A1=0,A2=0,A3=0,A4=0,A5=0,A6=0];
             [A1 = 0, A2 = 0, A3 = 0, A4 = 0, A5 = 0, A6 = 0]
/*                    Right hand side vector                        */
(C38) FOR I THRU 6 DO (
      P[I]:-EXPAND(SUBST(SUBLIST,WRB[I])),
    );
```

/* *Amplitudes of approximation functions* */

(C39) KK:GENMATRIX(K,6,6);

$$\begin{pmatrix} -\frac{224\,E}{75} & -\frac{608\,E}{225} & -\frac{736\,E}{1575} & -\frac{64\,E}{225} & \frac{128\,E}{675} & -\frac{64\,E}{525} \\ -\frac{608\,E}{225} & -\frac{11552\,E}{2625} & -\frac{3232\,E}{7875} & -\frac{64\,E}{225} & 0 & -\frac{64\,E}{525} \\ -\frac{736\,E}{1575} & -\frac{3232\,E}{7875} & -\frac{224\,E}{675} & \frac{64\,E}{1575} & -\frac{128\,E}{4725} & -\frac{64\,E}{4725} \\ -\frac{64\,E}{225} & -\frac{64\,E}{225} & \frac{64\,E}{1575} & -\frac{256\,E}{75} & -\frac{256\,E}{225} & -\frac{256\,E}{175} \\ \frac{128\,E}{675} & 0 & -\frac{128\,E}{4725} & -\frac{256\,E}{225} & -\frac{6656\,E}{7875} & -\frac{256\,E}{525} \\ -\frac{64\,E}{525} & -\frac{64\,E}{525} & -\frac{64\,E}{4725} & -\frac{256\,E}{175} & -\frac{256\,E}{525} & -\frac{5888\,E}{4725} \end{pmatrix}$$

(C40) PP:GENVECTOR(P,6);

$$\begin{pmatrix} -\frac{128\,E}{75} \\ -\frac{128\,E}{75} \\ -\frac{128\,E}{525} \\ 0 \\ 0 \\ 0 \end{pmatrix}$$

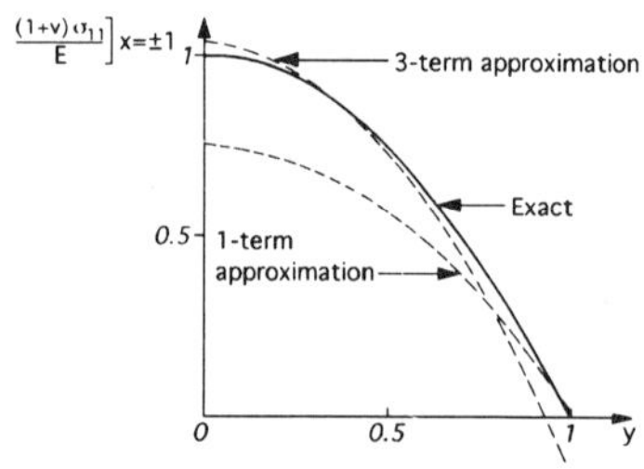

(C41) AA:KK^^-1 . PP ;

$$\begin{pmatrix} \frac{22016721}{36178867} \\ \frac{1770552}{36178867} \\ -\frac{8691081}{36178867} \\ -\frac{142497309}{723577340} \\ \frac{14558670}{36178867} \\ \frac{8691081}{723577340} \end{pmatrix}$$

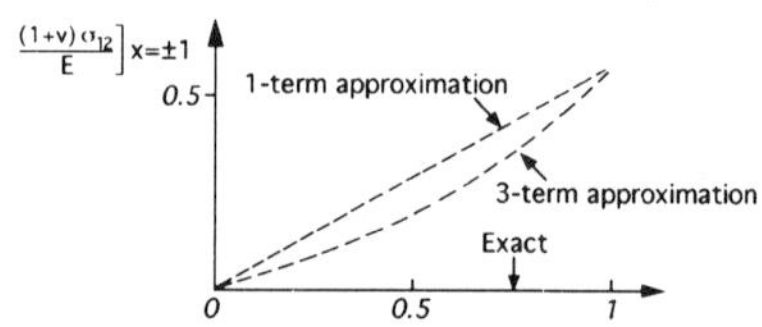

APPENDIX V - LIST OF BOOKS, MONOGRAPHS AND CONFERENCE PROCEEDINGS ON SYMBOLIC COMPUTATION

For the benefit of the readers of this monograph, all the monographs, textbooks and conference proceedings that are known to the author on symbolic computations are listed subsequently. The books and monographs are listed alphabetically. The conference proceedings are arranged in chronological order.

Books and Monographs

1. Abell, M.L. and Braselton, J.P.: Mathematica by Example, Academic Press, Boston, 1993.
2. Abell, M.L. and Braselton, J.P.: Differential Equations with Mathematica, Academic Press, Boston, 1993.
3. Abell, M.L.: The Mathematica Handbook, Academic Press, Boston, 1992.
4. Akritas, A.G.: Elements of Computer Algebra with Applications, Wiley, New York, 1989.
5. Beltzer, A.I.: Variational and Finite Element Methods - A Symbolic Computation Approach, Springer-Verlag, New York, 1990.
6. Blachman, N.: Mathematica, A Practical Approach, Prentice Hall, Englewood Cliffs, New Jersey, 1992.
7. Blachman, N.: Mathematica: Quick Reference, Version 2, Variable Symbols, Inc., 1992.
8. Braden, B., Krug, D.K., McCartney, P.W. and Wilkinson, S.: Discovering Calculus with Mathematica, John Wiley, New York, 1992.
9. Buchberger, B., Collins, G.E. and Loos, R. (Eds.): Computer Algebra: Symbolic and Algebraic Computation, Springer-Verlag, New York, 1983.
10. Char, B.W., Geddes, K.O., Gonnet, G.H., Monagan, M.B. and Watt, S.M.: First Leaves: A Tutorial Introduction to MAPLE, Watcom Publications, Waterloo, Ontario, Canada, 1988.
11. Char, B.W., Geddes, K.O., Gonnet, G.H., Monagan, M.B. and Watt, S.M.: First Leaves: A Tutorial Introduction to Maple V, Springer-Verlag, New York, 1992.
12. Chudnovsky, D.V. and Jenks, R.D. (Eds.): Computer Algebra, Lecture Notes in Pure and Applied Mathematics, Vol. 113, M. Dekker, New York, 1989.
13. Cohen, A.M. (Ed.): Computer Algebra in Industry, John Wiley, New York, 1993.
14. Crandall, R.E.: Mathematica for the Sciences, Addison-Wesley, Redwood City, CA, 1991.
15. Davenport, J.H., Siret, Y. and Tournier, E.: Computer Algebra: Systems and Algorithms for Algebraic Computation, Academic Press, San Diego, 1988.
16. DERIVE™, A Mathematical Assistant, Software Warehouse, Inc., Honolulu, HI, 1988.
17. Ellis, W. and Lodi, E.: A Tutorial Introduction to Mathematica, Brooks/Cole, Monterey, CA, 1991.
18. Finch, J.K. and Lehmann, M.: Exploring Calculus with Mathematica, Addison-Wesley, Redwood City, CA, 1992.
19. Ganzha, V.G., Rudenko, V.M. and Vorozhtsov, E.V. (Eds.): Computer Algebra and Its Applications to Mechanics, Nova Science Publishers, Commack, New York, 1993.
20. Gaylord, R.J., Kamin, S.N. and Wellin, P.R.: Introduction to Programming with Mathematica, Springer-Verlag, New York, 1993.

21. Gilligan, L. and Marquardt, J.F., Sr.: Calculus and the Derive(R) Program: Experiments with the Computer, MathWare, 604 East Mumford, Urbana, IL 61801, 1990.
22. Glynn, J.: Exploring Math From Algebra to Calculus with Derive(R) - A Mathematical Assistant for Your Personal Computer, MathWare, 604 East Munford, Urbana, IL 61801, 1989.
23. Gray, T. and Glynn, J.: Exploring Mathematics with Mathematica, Addison-Wesley, Redwood City, CA, 1991.
24. Gray, T. and Glynn, J.: The Beginner's Guide to Mathematica Version 2, Addison-Wesley, Redwood City, CA, 1992.
25. Heck, A.: The Maple Handbook, Springer-Verlag, New York, 1993.
26. Heck, A.: Introduction to Maple, Springer-Verlag, New York, 1993.
27. Heck, A.: Introduction to Maple: A Computer Algebra System, Springer-Verlag, New York, 1993.
28. Hearn, A.C.: REDUCE User's Manual, The Rand Corporation, 1987.
29. Hearn, A.C.: REDUCE Bibliography, The Rand Corporation, 1989.
30. Hirota, R. and Ito, M.: Introduction to REDUCE - Doing Symbolic Computation on PC, Science sha, Tokyo (in Japanese), 1989.
31. Hoft, M.H.: Laboratories for Calculus I Using Mathematica, Addison-Wesley, Redwood City, CA, 1992.
32. Howard, J.C.: Practical Applications of Symbolic Computations, IPC Science and Technology Press, Guildford, England, 1980.
33. Jenks, R.D. and Sutor, R.S.: Axiom: The Scientific Computation System, Springer-Verlag, New York, 1992.
34. Klimov, D.M. and Rudenko, V.M.: Methods of Computer Algebra in Mechanics Problems, Izdatel'stvo Nauka, Moscow (in Russian), 1989.
35. Knuth, D.E.: The Art of Computer Programming, Vol. 2: Semi-Numerical Algorithms, second edition, Addison-Wesley, Reading, MA, 1981.
36. Koike, S.: Mathematica: Introduction to Algebraic Computation, Gijutsuhyooronsha, Tokyo, Japan (in Japanese), 1990.
37. Kowalik, J.S. (Ed.): Coupling Symbolic and Numerical Computing in Expert Systems, North Holland, New York, 1986.
38. Kowalik, J.S. and Kitzmiller, C.T. (Eds.): Coupling Symbolic and Numerical Computing in Expert Systems II, North Holland, New York, 1988.
39. Lipson, J.D.: Algebra and Algebraic Computing, Addison-Wesley, Reading, MA, 1981.
40. MACSYMA Reference Manual, Vers. 11, Symbolics, Inc., Cambridge, MA, 1985.
41. MACSYMA User's Guide, Symbolics, Inc., Cambridge, MA, 1987.
42. Maeder, R.E.: Programming in Mathematica, Addison-Wesley, Redwood City, CA, 1991.
43. Mignotte, M.: Mathematiques Pour le Calcul Formel, Presses Universitaires de France, Paris, 1989.
44. Nakamura, H. and Matsui, S.: Symbolic computation in Structural Mechanics Using REDUCE, Gihodo Shuppan, Tokyo (in Japanese), 1989.
45. Ochiai, M. and Nagatomo, K.: Linear Algebra Using REDUCE, Kindai Kagaku sha, Tokyo (in Japanese), 1990.

46. Pavelle, R. (Ed.): Applications of Computer Algebra, Kluwer Academic Publishers, Boston, MA, 1985.
47. Rand, R.H.: Computer Algebra in Applied Mathematics: An Introduction to MACSYMA, Research Notes in Mathematics, Vol. 94, Pitman Publishing, Boston, MA, 1984.
48. Rand, R.H. and Armbruster, D.: Perturbation Methods, Bifurcation Theory and Computer Algebra, Springer-Verlag, New York, 1987.
49. Rayna, G.: REDUCE: Software for Algebraic Computation, Springer-Verlag, New York, 1987.
50. Redfern: The Maple Handbook, Springer-Verlag, New York, 1993.
51. Rehak, M.L., Dimaggio, F.L., Benaroya, H. and Elishakoff, I.: Random Vibrations with MACSYMA, Comput. Meth. Appl. Mech. and Eng., 61, 1987, 61-70.
52. Rice, J.R.: Mathematical Aspects of Scientific Software, IMA Volumes in Mathematics and Its Applications, Springer-Verlag, New York, 1988.
53. Skeel, R.D. and Keiper, J.B.: Elementary Numerical Computing with Mathematica, McGraw-Hill, New York, 1993.
54. Skiena, S.S.: Implementing Discrete Mathematics: Combinatories and Graph Theory with Mathematica, Addison-Wesley, Redwood City, CA, 1991.
55. Sims, C.C.: Abstract Algebra: A Computational Approach, John Wiley and Sons, New York, 1984.
56. Stauffer, D., Hehl, F.W., Winkelmann, V. and Zabolitzky, J.G.: Computer Simulation and Computer Algebra, Lectures for Beginners, Springer-Verlag, New York, 1989.
57. Stroyan, K.D.: Scientific Projects and Mathematical Background for Calculus Using Mathematica, Academic Press, Boston, 1993.
58. Symbolics, Bibliography of Publications Referencing MACSYMA: Computer Aided Mathematics Group, Eleven Cambridge Center, Cambridge, MA 02142, 1987.
59. Touretzky, D.S.: LISP: A Gentle Introduction to Symbolic Computation, Harper & Row, New York, 1984.
60. Tournier, E.: Computer Algebra and Differential Equations, Computational Mathematics and Applications, Academic Press, 1990.
61. Vardi, I.: Computational Recreations in Mathematica, Addison-Wesley, Redwood City, CA, 1991.
62. Vvedensky, D.: Partial Differential Equations with Mathematica, Addison-Wesley, Redwood City, CA, 1993.
63. Wade, E. and Lodi, E.: A Tutorial Introduction to Mathematica, Brooks/Cole, 1991.
64. Wagon, S.: Mathematica in Action, W.H. Freeman, San Francisco, CA, 1990.
65. Wickham-Jones: Computer Graphics with Mathematica, Springer-Verlag, New York.
66. Wolfram, S.: Mathematica: A System for Doing Mathematics by Computer, Addison-Wesley, Redwood City, CA, 1991.
67. Wolfram, S.: Mathematica Reference Guide, Addison-Wesley, Redwood City, CA, 1992.
68. Wooff, C. and Hodgkinson, D.: muMath: A Microcomputer Algebra System, Academic Press, London, 1987.
69. Zimmer, H.G.: Computational Problems, Methods and Results in Algebraic Number Theory, Lecture Notes in Mathematics, Vol. 268, Springer-Verlag, New York, 1972.

Conference Proceedings

1. Symbolic Languages in Data Processing, Proc. of a Symposium organized by Int. Computation Centre, Rome, March 26-31, 1962, Gordon & Breach Science, New York, 1962.
2. Petrick, S.R. (Ed.): Proc. Second Symposium on Symbolic and Algebraic Manipulation, March 23-25, 1971, Los Angeles, CA, Association for Computing Machinery, New York, 1971.
3. Jenks, R.D. (Ed.): Symsac '76: Proc. 1976 ACM Symposium on Symbolic and Algebraic Computation, Aug. 10-12, 1976, Yorktown Heights, New York, Association for Computing Machinery, New York, 1976.
3. Proc. 1977 MACSYMA User's Conference, University of California, Berkeley, July 27-29, 1977, NASA CP-2012.
4. Lewis, V.E. (Ed.): Proc. 1979 MACSYMA User's Conference, Washington, D.C., June 20-22, 1979.
5. Ng, E.W. (Ed.): Symbolic and Algebraic Computation: EUROSAM '79, International Symposium on Symbolic and Algebraic Manipulation, June 1979, Marseille, France, Springer-Verlag, Berlin, 1979.
6. Wang, P.S. (Ed.): Symsac '81: Proc. 1981 ACM Symposium on Symbolic and Algebraic Computation, Aug. 5-7, 1981, Snowbird, UT, Association for Computing Machinery, New York, 1981.
7. Calmet, J. (Ed.): Computer Algebra: EUROCAM 82 European Computer Algebra Conference, Marseille, France, April 5-7, 1982, Lecture Notes in Computer Science, No. 144, Springer-Verlag, Berlin, 1982.
8. Buchberger, B., Collins, G.E. and Loos, R.G.K. (Eds.): Computer Algebra: Symbolic and Algebraic Computation, second edition, Springer-Verlag, Wien, 1983.
9. van Hulzen, J.A. (Ed.): Computer Algebra: EUROCAL '83, European Computer Algebra Conference, London, England, March 28-30, 1983, Lecture Notes in Computer Science, No. 162, Springer-Verlag, Berlin, 1983.
10. Atkinson, M.A.: Computational Group Theory, Proc. London Mathematical Society Symposium on Computational Group Theory, Academic Press, San Diego, 1984.
11. Fitch, J. (Ed.): EUROSAM 84, Int. Symposium on Symbolic and Algebraic Computation, Proc. Symposium, July 9-11, 1984, Cambridge, England, Lecture Notes in Computer Science, No. 174, Springer-Verlag, 1984.
12. Golden, V.E. (Ed.): Proc. 1984 MACSYMA User's Conference, Schenectady, NY, July 23-25, 1984, General Electric, Schenectady, New York, 1984.
13. Hitotumatu, S. (chairman): Fifth RIMS Conference on Formula Manipulation and Its Application to Mathematical Study, Kyoto, Japan, 1985.
14. Inada, N. and Soma, T. (Eds.): Proc. of the Second RIKEN Int. Symposium on Symbolic and Algebraic Computation by Computers, Wako-shi, Japan, Series in Computer Science, Vol. 2, World Scientific Publishers, Philadelphia, PA, 1985.
15. Pavelle, R. (Ed.): Applications of Computer Algebra (papers based on the proc. Symposium on Symbolic Algebraic Manipulation in Scientific Computation presented by the ACS Division of Computers in Chemistry), 188th Meeting of the American Chemical Society, Aug. 26-31, 1984, Kluwer Academic Publishers, Boston, MA, 1985.
16. Carr, B.W. (Ed.): Symsac '86: Proc. 1986 Symposium on Symbolic and Algebraic Computation, July 21-23, 1986, Waterloo, Ontario, Canada, Association for Computing Machinery, New York, 1986.

17. Char, B.W. (Ed.): Symsac '86: Proc. 1986 ACM Symposium on Symbolic and Algebraic Computation, July 21-23, 1986, Waterloo, Ontario, Association for Computing Machinery, Baltimore, Maryland, 1986.
18. Caviness, B.F. (Ed.): EUROCAL '85, Proc. European Conference on Computer Algebra, April 1-3, 1985, Linz, Austria, Vol. 2, Research Contributions, Lecture Notes in Computer Science No. 203-204, Springer-Verlag, Berlin, 1986.
19. Buchberger, B. (Ed.): EUROCAL '85, Proc. European Conference on Computer Algebra, April 1-3, 1985, Linz, Austria, Vol. 1, Invited Lectures, Lecture Notes in Computer Science, No. 203-204, Springer-Verlag, Berlin, 1986.
20. Janssen, R. (Ed.): Trends in Computer Algebra, Proc. Int. Symposium, Bad Neuenahr, May 19-21, 1987, Lecture Notes in Computer Science, Vol. 296, Springer-Verlag, Berlin, 1987.
21. Bau, H.H., Herbert T. and Yovanovich, M.M. (Eds.): Symbolic Computations in Fluid Mechanics and Heat Transfer, Proc. symposium, ASME Winter Annual Meeting, Nov. 27-Dec. 2, 1988, Chicago, IL, HTD Vol. 105, AMD Vol. 97, ASME, New York, 1988.
22. Della Dora, J. and Fitch, J. (Eds.): CAP-88, Proc. Workshop on Computer Algebra and Parallelism, University of Grenoble, France, June 1988, Academic Press, London, 1988.
23. Gianni, P. (Ed.): Symbolic and Algebraic Computation: Int. Symposium ISSAC '88, Rome, July 4-8, 1988, Springer-Verlag, Berlin, 1988.
24. Miola, A. (Ed.): ISSAC '88, 1988 Int. Symposium on Symbolic and Algebraic Computation, Rome, Lecture Notes in Computer Science, No. 358, Springer-Verlag, 1988.
25. Davenport, J.H. (Ed.): EUROCAL '87, Proc. European Conference on Computer Algebra, June 1987, Leipzig, West Germany, Lecture Notes in Computer Science No. 378, Springer-Verlag, Berlin, 1989.
26. Grossman, R. (Ed.): Symbolic Computation: Applications to Scientific Computing, Proc. Conference on Symbolic Computation, Moffett Field, CA, Jan. 15-16, 1987, Society for Industrial and Applied Mathematics, Philadelphia, PA, Vol. 5, 1989.
27. Proc. ACM-SIGSAM 1989 Int. Symposium on Symbolic and Algebraic Computation, July 17-19, 1989, Portland, OR, ACM Press, 1989.
28. Boyle, A. and Caviness, B.F. (Eds.): Future Directions for Research in Symbolic Computation, Report of a Workshop on Symbolic Computation, April 29-30, 1988, Washington, D.C., Society for Industrial and Applied Mathematics, Philadelphia, PA, 1990.
29. ISSAC '90, Proc. Int. Symposium on Symbolic and Algebraic Computation, Tokyo, Japan, ACM Press/Addison Wesley, 1990.
30. Miola, C.A. (Ed.): DISCO '90, Proc. Symposium on Design and Implementation of Symbolic Computation Systems, April 10-12, 1990, Capri, Italy, Lecture Notes in Computer Science, Vol. 429, Springer-Verlag, Berlin, 1990.
31. Noor, A.K., Elishakoff, I. and Hulbert, G. (Eds.): Symbolic Computations and Their Impact on Mechanics, Proc. of an ASME Symposium, Nov. 25-30, 1990, Dallas, TX, PVP Vol. 205, ASME, New York, 1990.
32. Qingsheng, T.: Symbolic and Algebraic Manipulation for Formulae of Interpolation and Quadrature, ISSAC '90; Proc. Int. Symposium on Symbolic and Algebraic Computation, Aug. 20-24, 1990, Tokyo, Japan, Association for Computing Machinery, New York, 1990.

PERTURBATION METHODS AND COMPUTER ALGEBRA IN MECHANICS

R.H. Rand
Cornell University, Ithaca, NY, USA

OUTLINE

1. Introduction
2. Regular Perturbations: Mathieu's Equation
3. Padé Approximants
4. Method of Composite Expansions (Matched Aymptotics)
5. Method of Averaging: van der Pol's Equation
6. Hopf Bifurcations
7. Averaging with Elliptic Functions
8. Two Variable Expansion Method
9. Application to Coupled Oscillators

1. INTRODUCTION

Perturbation methods are a collection of diverse schemes for obtaining approximate analytic solutions to problems which involve a small parameter ϵ, usually in the form of a power series in ϵ. In these notes we will discuss computer algebra treatments of regular perturbations (section 2), and of three singular perturbation methods (composite expansions, section 4, averaging, section 5 and two variable expansion method, section 8).

Computer algebra has had a profound effect on the computations associated with perturbation methods [8,11]. Compared with hand calculations, computer algebra offers improvements in accuracy and speed, as well as in ease of computation. In particular, computer algebra has made major contributions to perturbation methods in the following areas:

a. Obtaining a large number of terms. This raises the question of the convergence of such series, and of the use of techniques for accelerating convergence [14], such as Pade approximants (section 3).

b. Obtaining solutions involving many unevaluated parameters. This permits the search for special relationships between the parameters, i.e. bifurcation theory. E.g. a curve in a parameter plane may separate systems into sets which exhibit qualitatively distinct behavior. A specific example is the Hopf bifurcation (section 6) which gives conditions under which a periodic motion called a limit cycle can be born in a system of nonlinear ODE's.

c. Manipulating special functions, such as Jacobian elliptic functions (section 7), which involve identities so complicated that hand calculation is difficult and discouraging.

2. REGULAR PERTURBATIONS: MATHIEU'S EQUATION

We begin with a mechanics problem: The linear stability of the upside-down pendulum under vertical forcing. Let θ be the angle between the pendulum and the upwards vertical direction. Then the governing linearized differential equation is

$$\frac{d^2\theta}{dt^2} - \left[\frac{g}{L} - \frac{A}{L}\cos t\right]\theta = 0 \tag{2.1}$$

where g is the acceleration of gravity, L is the pendulum's length, and A cos t is the displacement of the pendulum pivot above a fixed datum. The equilibrium at $\theta = 0$ will be said to be stable if all solutions to (2.1) are bounded, and unstable if an unbounded solution exists.

Eq.(2.1) is a special case of Mathieu's equation,

$$\frac{d^2x}{dt^2} + \left[\delta + \epsilon\cos t\right]x = 0 \tag{2.2}$$

in which $\delta = -\frac{g}{L}$ and $\epsilon = \frac{A}{L}$. As is well known [13], for small ϵ the stability chart in the δ–ϵ plane contains smooth "transition" curves separating regions of stability from regions of instability. Pairs of these curves emanate from the positive δ axis at $\delta = \frac{n^2}{4}$, n=1,2,3,..., in Arnold tongues of instability. In addition, there is a single transition curve which passes through $\delta = 0$, $\epsilon = 0$, tangent to the ϵ axis. Since for the forced pendulum we are interested in values of δ which are negative, it is the latter transition curve which we shall be interested in.

From Floquet theory [13] we learn that on these transition curves there exist periodic solutions of the period of the forcer, or of twice the period of the forcer, i.e., of period 2π or 4π. In order to obtain an approximate expression for the transition curve through $\delta = \epsilon = 0$, we shall expand δ and x in a power series in ϵ, a procedure called <u>regular perturbations</u>. (If ϵ appears in the approximate solution in any manner besides that of a power series, the procedure is called <u>singular perturbations</u>.) We set

(2.3) $$\delta = \delta_1 \epsilon + \delta_2 \epsilon^2 + O(\epsilon^3)$$

(2.4) $$x = x_0 + x_1 \epsilon + x_2 \epsilon^2 + O(\epsilon^3)$$

where δ_i are constant and x_i are functions of t. We substitute (2.3),(2.4) into (2.2), collect terms and equate to zero coefficients of ϵ^n for n = 0,1,2,... . This gives:

(2.5) $$\epsilon^0: \quad \ddot{x}_0 = 0$$

(2.6) $$\epsilon^1: \quad \ddot{x}_1 = - \delta_1 x_0 - x_0 \cos t$$

(2.7) $$\epsilon^2: \quad \ddot{x}_2 = - \delta_2 x_0 - \delta_1 x_1 - x_1 \cos t$$

From (2.5) we take $x_0(t) = 1$ for a periodic solution (which may be normalized to unity because eq.(2.2) is linear.) Substituting this into (2.6), we see that $x_1(t)$ will not be periodic unless $\delta_1 = 0$, whereupon $x_1(t) = \cos t$. Substituting all these results into (2.7) gives

(2.8) $$\ddot{x}_2 = - \delta_2 - \cos^2 t$$

Trigonometric reduction of (2.8) gives

$$\ddot{x}_2 = -\delta_2 - \frac{1}{2} - \frac{1}{2}\cos 2t \tag{2.9}$$

From (2.9) we see that $x_2(t)$ will not be periodic unless $\delta_2 = -\frac{1}{2}$. Thus we have derived the following approximate expression for the desired transition curve:

$$\delta = -\frac{1}{2}\epsilon^2 + O(\epsilon^3) \tag{2.10}$$

Returning to the upside-down pendulum, eq.(2.1), we see that for small values of g/L, the pendulum can be made stable by choosing A/L appropriately.

By using computer algebra we can formalize this process and thereby obtain many terms of the foregoing series:

$$\delta = -\frac{\epsilon^2}{2} + \frac{7\,\epsilon^4}{32} - \frac{29\,\epsilon^6}{144} + \frac{68687\,\epsilon^8}{294912} - \frac{123707\,\epsilon^{10}}{409600} \tag{2.11}$$

$$+ \frac{8022167579\,\epsilon^{12}}{19110297600} - \frac{286241141477\,\epsilon^{14}}{468202291200} + \frac{7534554811777337\,\epsilon^{16}}{8182428094955520}$$

$$- \frac{6364218991597629688 7\,\epsilon^{18}}{447374256091693056 00} + \frac{40116328088292198921753 01\,\epsilon^{20}}{17894970243667722240000 00}$$

$$- \frac{129537384934904738246595677\,\epsilon^{22}}{36088189991396573184000000}$$

$$+ \frac{4183277971077047625062883425563 63\,\epsilon^{24}}{71839171935513536788168704000000}$$

$$- \frac{231905723577803781627393585040227\,1\,\epsilon^{26}}{2428164011420357543440102195200000} + \frac{780318809060771190938074412048864142\,07\,\epsilon^{28}}{493546818321293414755528920268800000 0} - \frac{988787251996908762543840889495263447522476\,3\,\epsilon^{30}}{374787115162732186829979773829120000000000\,0} + \ldots$$

Next we present the record of a **MACSYMA** run in which the steps which we accomplished by hand to $O(\epsilon^2)$ above are repeated using computer algebra. This run is accomplished in an interactive mode, i.e., the instructions are typed from the keyboard. After this run we will present a program which automates the process.

```
(C4) X0:1;
(D4)                              1
(C5) DELTA:DELTA1*E+DELTA2*E^2;

                                  2
(D5)                    DELTA2 E   + DELTA1 E

(C6) DEPENDS([X1,X2],T);
(D6)                        [X1(T), X2(T)]
(C7) X:X0+X1*E+X2*E^2;
                              2
(D7)                         E   X2 + E X1 + 1

(C8) DIFF(X,T,2)+(DELTA+E*COS(T))*X;

            2
         2 d X2                            2
(D8) E   ---- + (E COS(T) + DELTA2 E   + DELTA1 E)
             2
           dT
                                                          2
                                           2             d X1
                                        (E   X2 + E X1 + 1) + E ----
                                                                  2
                                                                dT
```

```
(C9) DE:TAYLOR(%,E,0,2);

                2
               d X1
(D9)/T/       (----- + COS(T) + DELTA1) E
                  2
               dT

                    2
                   d X2                                        2
               + (----- + (COS(T) + DELTA1) X1 + DELTA2) E  + . . .
                      2
                   dT

(C10) EQ1:COEFF(DE,E);

                            2
                           d X1
(D10)/R/                   ----- + COS(T) + DELTA1
                              2
                           dT

(C11) EQ2:COEFF(DE,E,2);

                     2
                    d X2
(D11)/R/            ----- + (COS(T) + DELTA1) X1 + DELTA2
                       2
                    dT

(C12) DELTA1:0;
(D12)                                   0
(C13) EV(EQ1);
                                  2
                                 d X1
(D13)/R/                         ----- + COS(T)
                                    2
                                 dT

(C14) ODE2(%,X1,T);
(D14)                    X1 = COS(T) + %K2 T + %K1

(C15) X1:RHS(%),%K1:%K2:0;
(D15)                                 COS(T)
(C16) EV(EQ2);
                           2
                          d X2       2
(D16)/R/                  ----- + COS (T) + DELTA2
                             2
                          dT
```

```
(C17) EXPAND(TRIGREDUCE(EXPAND(%)));
```

$$\text{(D17)} \qquad \frac{d^2 X2}{dT^2} + \frac{COS(2\ T)}{2} + DELTA2 + \frac{1}{2}$$

```
(C18) DELTA2:-1/2;
```

$$\text{(D18)} \qquad -\frac{1}{2}$$

```
(C19) EV(DELTA);
```

$$\text{(D19)} \qquad -\frac{E^2}{2}$$

One advantage of converting the above run to a program is that the order of truncation may be controlled by assigning a variable (called "trunc" below) which adjusts the size of sums and DO loops. Here is an outline of how the program works, together with the corresponding lines in the above run:

1. Set up expansions (C4–C7)
2. Plug into D.E. and collect terms (C8–C11)

DO LOOP:

3. Apply trig identities (C17)
4. Pick off "secular" terms (which produce non-periodic solutions) and find δ_i (C18)
5. Use O.D.E. package to solve for x_i (C12–C16)

END OF DO LOOP

6. Output results (C19)

Here is the program:

```
/* set up expansions */
trunc:read("truncation order")$
depends(x,t);
xx:sum(x[i]*e^i,i,0,trunc);
delta:sum(del[i]*e^i,i,0,trunc);
results:[x[0]=1,del[0]=0];
/* prepare null function for later */
null(t):=0;

/* plug into d.e. and collect terms */
de1:diff(xx,t,2)+(delta+e*cos(t))*xx$
de2:taylor(de1,e,0,trunc)$
for i:0 thru trunc do eq[i]:coeff(de2,e,i)$

/* main loop */
for i:1 thru trunc do (
temp1:ev(eq[i],results),

/* apply trig identities */
temp2:expand(trigreduce(expand(temp1))),

/* pick off secular terms */
temp3:subst(null,cos,temp2),
temp4:ev(temp3,x[i]=0),

/* solve for del[i] */
temp5:solve(temp4,del[i]),
results:append(results,temp5),
temp6:ev(temp2,results),
temp7:ratsimp(temp6),

/* use o.d.e. package to find x[i] */
temp8:ode2(temp7,x[i],t),
/* zap arbitrary constants */
temp9:ev(temp8,%k1=0,%k2=0),
results:append(results,[temp9]))$
/* end of main loop */

/* output results */
deltafinal:ev(delta,results);
```

And here is a sample run:

```
(C3) /* set up expansions */
trunc:read("truncation order")$
truncation order
8;

(C4) depends(x,t);

(D4)                [X(T)]

(C5) xx:sum(x[i]*e^i,i,0,trunc);

             8       7       6       5
(D5) X  E  + X  E  + X  E  + X  E
      8       7       6       5

            4       3       2
     + X  E  + X  E  + X  E  + X  E + X
        4       3       2       1     0

(C6) delta:sum(del[i]*e^i,i,0,trunc);

      8         7         6
(D6) E  DEL  + E  DEL  + E  DEL
           8         7         6

   5         4         3         2
 + E  DEL  + E  DEL  + E  DEL  + E  DEL
         5         4         3         2

 + E DEL  + DEL
        1      0

(C7) results:[x[0]=1,del[0]=0];

(D7)        [X  = 1, DEL  = 0]
              0          0

JUMP TO END OF RUN...

/* output results */
deltafinal:ev(delta,results);

                 8       6      4   2
         68687 E    29 E    7 E    E
(D13)    -------- - ----- + ---- - --
          294912     144     32    2
```

3. PADE APPROXIMANTS

Pade approximants are a tool for improving the convergence of perturbation series [1]. Given a truncated power series of the form

$$f = a_0 + a_1 \epsilon + a_2 \epsilon^2 + \cdots + a_n \epsilon^n \tag{3.1}$$

a Pade approximant is a <u>rational</u> <u>fraction</u> which has the same truncated Taylor series as f up to $O(\epsilon^n)$. For example, the truncated series

$$1 + \frac{\epsilon}{2} - \frac{\epsilon^2}{8} \quad \text{has the Pade approximant} \quad \frac{3\epsilon + 4}{\epsilon + 4} \tag{3.2}$$

since a Taylor expansion gives

$$\frac{3\epsilon + 4}{\epsilon + 4} = 1 + \frac{\epsilon}{2} - \frac{\epsilon^2}{8} + \frac{\epsilon^3}{32} - \frac{\epsilon^4}{128} + \cdots \tag{3.3}$$

The process of finding a Pade approximant involves solving a system of linear equations [1], but **MACSYMA** can do this for you automatically.

As a model example, suppose that the function $\sqrt{1+\epsilon}$ plays the role of an exact solution to a perturbation problem. We normally do not have access to the exact solution, but rather we solve for its perturbation expansion, the role of which in this case is played by the following Taylor series:

$$\sqrt{1+\epsilon} = 1 + \frac{\epsilon}{2} - \frac{\epsilon^2}{8} + \frac{\epsilon^3}{16} - \frac{5}{128}\epsilon^4 + \frac{7}{256}\epsilon^5 - \frac{21}{1024}\epsilon^6 + \cdots \tag{3.4}$$

Now this series diverges for $|\epsilon|>1$ (since $\sqrt{1+\epsilon}$ has a branch point at $\epsilon = -1$). For example, at $\epsilon = 2$, the "exact solution" gives $\sqrt{1+\epsilon} = \sqrt{3} = 1.7320$, while the "perturbation solution" gives

$1+\frac{\epsilon}{2}-\cdots-\frac{21}{1024}\epsilon^6=0.9375$. Now we attempt to improve the convergence of the series by using a Pade approximant. In **MACSYMA**, the command is

(3.5) PADE (taylor series, m, n);

where m is the degree of the polynomial in the numerator of the Pade approximant, and n is the corresponding value for the denominator. These should be chosen so that m+n equals the degree of the truncated Taylor series. In this example we take m = n = 3 and obtain:

```
(C4) sqrt(1+e);

(D4)            SQRT(E + 1)

(C5) taylor(%,e,0,6);

                    2     3       4       5        6
              E    E     E     5 E     7 E     21 E
(D5)/T/ 1 +  - -  -- +  -- -  ---- +  ---- -  ----- + . . .
              2    8    16     128     256     1024

(C6) pade(%,3,3);

             3       2
          7 E  + 56 E  + 112 E + 64
(D6)   [---------------------------]
            3      2
           E  + 24 E  + 80 E + 64

(C7) [D4,D5,D6],e = 2,numer;

(D7)   [1.73205, 0.9375, [1.73171]]
```

That is, at ϵ = 2 the Pade approximant gives 1.73171, and thus very nearly recovers the exact solution from the divergent perturbation series!

Besides offering an improved approximation, there is another important use for Pade approximants. The <u>zeros</u> of the <u>denominator</u> of the Pade approximant often give useful information about the <u>location</u> and nature of the nearest <u>singularity</u> in the perturbation solution. To continue with the above model example, the zeros of the denominator of the Pade approximant in (D6), $\epsilon^3+24\epsilon^2+80\epsilon+64=0$, all lie on the negative ϵ axis:

```
(C8) DENOM(PART(D6,1));

            3       2
(D8)       E  + 24 E  + 80 E + 64

(C9) ALLROOTS(%);

(D9) [E = - 1.23191, E = - 2.57242, E = - 20.19566]
```

It turns out that the closest zero to the origin ($\epsilon=0$) approaches the branch point $\epsilon=-1$ as the number of terms increases.

My experience is that Pade approximants often give excellent results in real perturbation problems. As an example, we take the transition curve in the Mathieu equation that we found at the end of the previous section. Numerical integration of the differential equation (2.2) shows that for $\delta = -0.4$, for example, the transition curve passes close to $\epsilon = 1.035$. Substituting this value of ϵ into the $O(\epsilon^8)$ perturbation expansion for the transition curve gives

$$\delta = -\frac{\epsilon^2}{2} + \frac{7\epsilon^4}{32} - \frac{29\epsilon^6}{144} + \frac{68687\epsilon^8}{294912} = -0.225 \text{ for } \epsilon = 1.035$$

This value is far from the numerically-obtained value of $\delta = -0.4$. Now we use Pade approximants:

```
(C7) TC;
             2      4       6          8
            E    7 E    29 E    68687 E
(D7)/T/ - -- + ---- - ----- + -------- + . . .
            2    32     144    294912
```

```
(C8) PADE(TC,4,4);

(D8)
                       4              2
          2092824 E  + 2244096 E
   [- ─────────────────────────────────]
                4              2
       882545 E  + 6149232 E  + 4488192
```

which gives the value $\delta = -0.398$ for $\epsilon = 1.035$, which is very close to the numerical value of $\delta = -0.4$.

We may also use the Pade approximant to estimate the radius of convergence of the perturbation power series:

```
(C9) DENOM(PART(D8,1));

                4              2
(D9) 882545 E  + 6149232 E  + 4488192

(C10) ALLROOTS(%);

(D10) [E = 0.91014 %I, E = - 0.91014 %I,

       E = 2.47775 %I, E = - 2.47775 %I]
```

The closest root of the denominator of the Pade approximant to the expansion point ϵ=0 is a distance 0.91 away. Thus we expect a radius of convergence in ϵ of about 0.9. Taking more terms gives a value of about 0.73. Thus the value we used above, namely $\epsilon = 1.035$, was outside the radius of convergence, which explains why the perturbation series did not deliver a meaningful value. However, note how well the Pade approximant improved the convergence of this divergent series.

4. METHOD OF COMPOSITE EXPANSIONS (MATCHED ASYMPTOTICS) [9]

The method of composite expansions is a singular perturbation method which is an alternative to matched aymptotic expansions for differential equations of the form:

$$\epsilon\, y'' + a(x)\, y' + b(x)\, y = 0 \tag{4.1}$$

with two boundary conditions:

$$y(0) = \alpha, \quad y(1) = \beta, \tag{4.2}$$

where primes represent differentiation with respect to x. In (4.1), $\epsilon << 1$ and a(x) and b(x) are analytic functions of x on [0,1]. We assume that $a(x) > 0$ throughout [0,1], in which case a boundary layer (b.l.) will occur at $x = 0$. (In the opposite case that $a(x) < 0$, the b.l. will occur at $x = 1$, but changing independent variable to $\xi = 1-x$ will bring the b.l. to $\xi = 0$. We assume that a(x) does not change sign in [0,1], since this case leads to more complicated behavior (internal b.l.'s), see [1,7].)

The following description of the method is based on Nayfeh [7], pp.148–149: We look for a solution to (4.1) and (4.2) in the form:

$$y = \sum_{n=0}^{\infty} \epsilon^n f_n(x) + e^{-g(x)/\epsilon} \sum_{n=0}^{\infty} \epsilon^n h_n(x) \tag{4.3}$$

in which the functions f_n, g and h_n are to be found. We require $g(x) \sim x$ in the limit $x \to 0$, since the b.l. is known to have thickness x. The procedure is to substitute (4.3) into (4.1) and (4.2), collect terms, and equate the coefficients of ϵ^n and $\epsilon^n e^{-g/\epsilon}$ to zero. We proceed directly to an example [3]:

(4.4) $$\epsilon\, y'' + (2x+1)\, y' + 2\, y = 0$$

for which $a(x) = 2x+1$ and $b(x) = 2$. Substituting (4.3) into (4.4) and multiplying by ϵ gives the following lowest order terms:

(4.5) $$e^{-g/\epsilon}: \qquad g'\, h_0\, (g' - 2x - 1) = 0$$

(4.6) $$\epsilon\, e^{-g/\epsilon}: \quad h_0\, (2 - g'') + h_0'\, (2x + 1 - 2g') = 0$$

(4.7) $$\epsilon: \qquad 2\, f_0 + (2x + 1)\, f_0' = 0$$

Substituting (4.3) into the b.c. (4.2) gives:

(4.8) $$f_0(0) + h_0(0) = \alpha \quad \text{and} \quad f_0(1) = \beta$$

Eq.(4.5) gives either $h_0 = 0$, g = constant (both of which must be rejected), or:

(4.9) $$g(x) = x^2 + x$$

in which the arbitrary constant of integration has been taken as zero in order that $g(x) \sim x$. Substituting (4.9) into (4.6) gives

(4.10) $$h_0' = 0 \quad \text{or} \quad h_0(x) = \text{constant}.$$

Next, (4.7) may be integrated to give:

(4.11) $$f_0(x) = \frac{\text{constant}}{2x + 1}.$$

The arbitrary constants of integration in (4.10) and (4.11) are found by using the b.c.(4.8):

$$f_0(x) = \frac{3\beta}{2x + 1}, \quad h_0(x) = \alpha - 3\beta \tag{4.12}$$

Substituting (4.9) and (4.12) into (4.3) gives the approximate result:

$$y = \frac{3\beta}{2x + 1} + (\alpha - 3\beta)\, e^{-\frac{x^2+x}{\epsilon}} + O(\epsilon) \tag{4.13}$$

This type of computation may be automated using computer algebra. The **MACSYMA** program "composite" which we present here accomplishes this task for general functions a(x) and b(x), to arbitrary order of truncation. Here is a sample run on the previous example:

```
composite();
The d.e. is: ey''+a(x)y'+b(x)y=0
with b.c. y(0)=y0 and y(1)=y1
enter a(x) > 0 on [0,1]
2*x+1;
enter b(x)
2;
enter y0
alpha;
enter y1
beta;

The d.e. is: ey''+( 2 x + 1 )y'+( 2 )y=0
with b.c. y(0)= alpha and y(1)= beta

enter truncation order
3;

                  3                2
       85312 beta e     928 beta e     16 beta e
     (- ------------- - ------------ - ----------
            243              27            3

                                2
                               x  + x
                             - ------
                                 e
 - 3 beta + alpha) %e

   3            6                5                4
 - e  (5120 beta x  + 15360 beta x  + 21504 beta x
```

$$+ 17408\,\beta x^3 + 16032\,\beta x^2 + 9888\,\beta x$$

$$- 85312\,\beta)/(31104\,x^7 + 108864\,x^6 + 163296\,x^5$$

$$+ 136080\,x^4 + 68040\,x^3 + 20412\,x^2 + 3402\,x + 243)$$

$$- e^2\,(128\,\beta x^4 + 256\,\beta x^3 + 336\,\beta x^2$$

$$+ 208\,\beta x - 928\,\beta)/(864\,x^5 + 2160\,x^4 + 2160\,x^3$$

$$+ 1080\,x^2 + 270\,x + 27)$$

$$- \frac{e\,(8\,\beta x^2 + 8\,\beta x - 16\,\beta)}{24\,x^3 + 36\,x^2 + 18\,x + 3} + \frac{3\,\beta}{2\,x + 1}$$

In order to check the accuracy of the results, we used finite differences to obtain a numerical solution to (4.4),(4.2) for $\alpha = 1$, $\beta = 2$. The following table lists the values of $y'(0)$ obtained from various truncations of the composite expansion, along with the finite difference value, for $\epsilon = 0.01, 0.05$ and 0.09:

Table of approximate values of $y'(0)$

truncation order	$\epsilon = 0.01$	$\epsilon = 0.05$	$\epsilon = 0.09$
0	488	88.0	43.55
1	497.97	95.2	47.98
2	498.59	96.88	48.48
3	498.65	97.40	46.94
finite diff.	498.66	98.64	54.16

Note that while additional accuracy is obtained by taking more terms in the expansion for $\epsilon = 0.01$ and $\epsilon = 0.05$, this is no longer true for $\epsilon = 0.09$. This behavior is typical of asymptotic series (as opposed to convergent series.) The two limiting processes of letting $\epsilon \to 0$ and truncating the series after $N \to \infty$ terms are in competition. In an asymptotic series, we fix N and let $\epsilon \to 0$, while in a convergent series, we fix ϵ and let $N \to \infty$. In terms of the Table, if we fix the truncation order at N = 3 and take ϵ smaller, the difference between the asymptotic series approximation and the exact solution (represented by the finite differences result) becomes smaller. If we fix ϵ, however, we cannot expect that increasing the order of truncation N will necessarily increase the accuracy of the approximation, cf. $\epsilon = 0.09$.

Here is the program listing:

```
/* method of composite expansions */

composite():=(

/* input problem from keyboard */
print("The d.e. is: ey''+a(x)y'+b(x)y=0"),
print("with b.c. y(0)=y0 and y(1)=y1"),
a:read("enter a(x) > 0 on [0,1]"),
b:read("enter b(x)"),
y0:read("enter y0"),
y1:read("enter y1"),
print("The d.e. is: ey''+(",a,")y'+(",b,")y=0"),
print("with b.c. y(0)=",y0,"and y(1)=",y1),
trunc:read("enter truncation order"),

/* set up basic form of solution */
y:sum(f[n](x)*e^n,n,0,trunc)+
  %e^(-g(x)/e)*sum(h[n](x)*e^n,n,0,trunc),

/* substitute into d.e. */
de1:e*diff(y,x,2)+a*diff(y,x)+b*y,

/* expand and collect terms */
de2:expand(e*de1),
gstuff:coeff(de2,%e^(-g(x)/e)),
other:expand(de2-gstuff*%e^(-g(x)/e)),
gstuff2:taylor(gstuff,e,0,trunc+1),
other2:taylor(other,e,0,trunc+1),
for i:0 thru trunc+1 do
   geq[i]:coeff(gstuff2,e,i),
for i:1 thru trunc+1 do
   otheq[i]:coeff(other2,e,i),
```

```
/* find g(x) */
geq:geq[0]/h[0](x)/diff(g(x),x),
gsol:ode2(geq,g(x),x),
resultlist:[g(x)=ev(rhs(gsol),%c=0)],

/* find h[i](x)'s */
/* hh[i] = constant of integration */
for i:0 thru trunc do(
   heq[i]:ev(geq[i+1],resultlist,diff),
   hsol[i]:ode2(heq[i],h[i](x),x),
   hsol[i]:ev(hsol[i],%c=hh[i]),
   resultlist:append(resultlist,[hsol[i]])),

/* find f[i](x)'s */
/* ff[i] = constant of integration */
for i:0 thru trunc do(
   feq[i]:ev(otheq[i+1],resultlist,diff),
   fsol[i]:ode2(feq[i],f[i](x),x),
   fsol[i]:ev(fsol[i],%c=ff[i]),
   resultlist:append(resultlist,[fsol[i]])),

/* boundary conditions */
/* y(0)=y0 and y(1)=y1 */
bc1:ev(y=y0,resultlist,x=0),
bc2:ev(y=y1,resultlist,x=1),
/* kill exponentially small %e terms */
bc2:subst(0,%e,bc2),
for i:0 thru trunc do
   (bc1[i]:ratcoef(bc1,e,i),
    bc2[i]:ratcoef(bc2,e,i)),

/* solve for unknown consts */
bceqs:makelist(bc1[i],i,0,trunc),
bceqs:append(bceqs,makelist(bc2[i],i,0,trunc)),
bcunk:makelist(hh[i],i,0,trunc),
bcunk:append(bcunk,makelist(ff[i],i,0,trunc)),
const:solve(bceqs,bcunk),

/* substitute back */
resultlist:ratsimp(ev(resultlist,const)),
ysol:ev(y,resultlist))$
```

5. METHOD OF AVERAGING: VAN DER POL'S EQUATION

The method of averaging [12] is a powerful perturbation method based on choosing a near-identity transformation to simplify the approximate equations of motion as much as possible. In contrast to regular perturbations, where the solution is expanded in a power series in ϵ, in averaging it is the near-identity transformation which is expanded in a power series in ϵ. The result of the method is a simplified differential equation rather than an approximate expression for the solution of the differential equation.

We shall use the method of averaging to treat systems of the form:

$$\ddot{z} + z = \epsilon\, F(z,\dot{z}) \tag{5.1}$$

The first step is to assume a solution in the form

$$z = a \cos \varphi\ , \qquad \dot{z} = - a \sin \varphi \tag{5.2}$$

in which a=a(t) and $\varphi=\varphi(t)$, i.e., the method of variation of parameters. Substituting (5.2) into (5.1) gives

$$\dot{a} = - \epsilon\, s\, F(a\, c,-a\, s)\ , \quad \dot{\varphi} = 1 - \frac{\epsilon}{a}\, c\, F(a\, c,-a\, s) \tag{5.3}$$

where $s = \sin \varphi$, $c = \cos \varphi$.

Next we posit a near-identity transformation of the form:

$$a = \bar{a} + \epsilon\, w_1 + \epsilon^2\, v_1 + \epsilon^3\, u_1 + O(\epsilon^4), \tag{5.4}$$

$$\varphi = \bar{\varphi} + \epsilon\, w_2 + \epsilon^2\, v_2 + \epsilon^3\, u_2 + O(\epsilon^4) \tag{5.5}$$

where the generating functions w_i, v_i and u_i depend on $\bar{a}$, $\bar{\varphi}$. We must substitute eqs.(5.4),(5.5) into eq.(5.3), expand in ϵ, collect terms, and equate to zero the coefficient of ϵ^n, for n=0,1,2,... . Then our goal is to choose the generating functions so as to simplify the resulting differential eqs. on $\bar{a}$ and $\bar{\varphi}$.

We shall illustrate the method by applying it to a particular example, van der Pol's equation:

$$\ddot{z} + z = \epsilon\, \dot{z}\, (1-z^2) \tag{5.6}$$

which is in the form of (5.1) with $F(z,\dot{z},t) = \dot{z}\,(1-z^2)$. This d.e. exhibits a periodic solution called a limit cycle. Making the substitution (2) we obtain the following differential eqs.(5.3), which are exact so far:

$$\dot{a} = \epsilon\, a\, s^2(1-a^2c^2)\ , \quad \dot{\varphi} = 1 + \epsilon\, c\, s(1-a^2c^2) \tag{5.7}$$

Substituting (5.4),(5.5) into (5.7), and temporarily neglecting terms of $O(\epsilon^2)$, gives

$$\dot{\bar{a}} + \epsilon\frac{\partial w_1}{\partial \bar{a}}\,\dot{\bar{a}} + \epsilon\frac{\partial w_1}{\partial \bar{\varphi}}\,\dot{\bar{\varphi}} = \epsilon\, \bar{a}\, \sin^2\bar{\varphi}\ [1-\bar{a}^2\cos^2\bar{\varphi}] + O(\epsilon^2) \tag{5.8}$$

and a similar equation on $\dot{\bar{\varphi}}$. Solving for $\dot{\bar{a}}$ and $\dot{\bar{\varphi}}$ gives

$$\dot{\bar{a}} = -\,\epsilon\frac{\partial w_1}{\partial \bar{\varphi}} + \epsilon\, \bar{a}\, \sin^2\bar{\varphi}\ [1-\bar{a}^2\cos^2\bar{\varphi}] + O(\epsilon^2) \tag{5.9}$$

$$\dot{\bar{\varphi}} = 1 - \epsilon\frac{\partial w_2}{\partial \bar{\varphi}} + \epsilon\, \sin\bar{\varphi}\, \cos\bar{\varphi}\ [1-\bar{a}^2\cos^2\bar{\varphi}] + O(\epsilon^2) \tag{5.10}$$

Trigonometric reduction of eqs.(5.9),(5.10) gives

$$\dot{\bar{a}} = -\epsilon\frac{\partial w_1}{\partial\bar{\varphi}} + \epsilon\,\bar{a}\left[\frac{1}{2} - \frac{\bar{a}^2}{8} - \frac{1}{2}\cos 2\bar{\varphi} + \frac{\bar{a}^2}{8}\cos 4\bar{\varphi}\right] + O(\epsilon^2) \tag{5.11}$$

and a similar eq. on $\dot{\bar{\varphi}}$. Now we may choose w_1 and w_2 to eliminate the trigonometric terms. From (5.11) this amounts to integrating the trig terms in $\bar{\varphi}$, holding $\bar{a}$ constant (which gives the same result as averaging these terms over one period, hence the name method of averaging):

$$w_1 = \int \bar{a}\left[-\frac{1}{2}\cos 2\bar{\varphi} + \frac{\bar{a}^2}{8}\cos 4\bar{\varphi}\right] d\bar{\varphi} \tag{5.12}$$

$$w_1 = -\frac{\bar{a}}{4}\sin 2\bar{\varphi} + \frac{\bar{a}^3}{32}\sin 4\bar{\varphi} + K_1 \tag{5.13}$$

where K_1 is an arbitrary function of $\bar{a}$. Similarly we find

$$w_2 = \left[-\frac{1}{4} + \frac{\bar{a}^2}{8}\right]\cos 2\bar{\varphi} + \frac{\bar{a}^2}{32}\cos 4\bar{\varphi} + K_2 \tag{5.14}$$

These choices for w_1 and w_2 give the following simplified eqs. on $\dot{\bar{a}}$ and $\dot{\bar{\varphi}}$:

$$\dot{\bar{a}} = \epsilon\,\bar{a}\left[\frac{1}{2} - \frac{\bar{a}^2}{8}\right] + O(\epsilon^2)\ , \qquad \dot{\bar{\varphi}} = 1 + O(\epsilon^2) \tag{5.15}$$

The limit cycle in van der Pol's eq.(5.6) corresponds to the non-zero equilibrium of the $\dot{\bar{a}}$ eq.(5.15), namely $\bar{a} = 2 + O(\epsilon)$.

The process thus far is called first-order averaging. To extend the method to second-order averaging, we again substitute (5.4),(5.5) into (5.7), but this time retain terms of $O(\epsilon^2)$, and use the expressions (5.13),(5.14) for $w_1 . w_2$. This will yield eqs. on $\dot{\bar{a}}$ and $\dot{\bar{\varphi}}$ which may be simplified by choosing v_1 and v_2 to kill any trig terms, as was done for w_1 and w_2 above. The process is then continued to third-order averaging in which u_1 and u_2 are similarly determined, and so on. As a time-saving short-cut, note that for the final round of averaging, we may avoid solving for the associated generating functions. E.g., for third-order averaging, the functions u_1 and u_2 need not actually be computed, but rather may be taken into account by simply removing the $O(\epsilon^3)$ periodic terms.

Here is a MACSYMA program which automates the foregoing computation:

```
/* third order averaging on vdp's eq */
depends([a,phi],t)$
z:a*cos(phi)$
zd:-a*sin(phi)$
f:(1-z^2)*zd$
de1:[diff(a,t)=-e*sin(phi)*f,diff(phi,t)=1-e/a*cos(phi)*f]$
depends([w1,w2,v1,v2],[abar,phibar])$
transf:[a=abar+e*w1+e^2*v1,phi=phibar+e*w2+e^2*v2]$
depends([abar,phibar],t)$
de2:de1,transf,diff$
de3:solve(%,[diff(abar,t),diff(phibar,t)])$
de4:taylor(de3,e,0,3)$

/* first order averaging: */
ws1:ratcoef(%,e)$
expand(trigreduce(expand(%)))$
integrate(%,phibar)$
ws2:%,%integconst1:k1,%integconst2:k2$
ws3:ws2-ratcoef(ws2,phibar)*phibar$
ws4:solve(part(%,1),[w1,w2])$
de5:de4,ws4,diff$
```

```
/* second order averaging: */
vs1:ratcoef(%,e,2)$
de6:expand(trigreduce(expand(%)))$
integrate(%,phibar)$
vs2:%,%integconst3:k3,%integconst4:k4$
vs3:vs2-ratcoef(vs2,phibar)*phibar$
vs4:solve(part(%,1),[v1,v2])$
de7:de5,vs4,diff$

/* third order averaging: */
de8:expand(trigreduce(expand(%)))$
subst(s,sin,%)$
de9:subst(c,cos,%)$
c(any):=0$
s(any):=0$
de10:ev(de9);
```

The result of this program is the following simplified ("averaged") d.e.'s on $\dot{\bar{a}}$ and $\dot{\bar{\varphi}}$:

$$\frac{dABAR}{dT} = -\frac{3\,ABAR^2\,E^3\,K3}{8} + \frac{E^3\,K3}{2} - \frac{3\,ABAR\,E^3\,K1^2}{8} - \frac{3\,ABAR^2\,E^2\,K1}{8} + \frac{E^2\,K1}{2} - \frac{43\,ABAR^7\,E^3}{16384} + \frac{39\,ABAR^5\,E^3}{2048} - \frac{9\,ABAR^3\,E^3}{256} - \frac{ABAR^3\,E}{8} + \frac{ABAR\,E}{2}, \tag{5.16}$$

$$\frac{dPHIBAR}{dT} = -\frac{11\,ABAR^3\,E^3\,K1}{64} + \frac{3\,ABAR\,E^3\,K1}{8} - \frac{11\,ABAR^4\,E^2}{256} + \frac{3\,ABAR^2\,E^2}{16} - \frac{E^2}{8} + 1$$

Note that the form of the averaged d.e.'s depends on the choice of the arbitrary functions of $\bar{a}$, K_1 and K_3. This interesting point has been discussed by Kahn and his associates (see e.g.[6]), who show that there are various ways to select the constants K_i. For example, if we take all K_i=0, then we achieve the simplest choice for the generating functions (which relate the transformed coordinates to the original coordinates). However, a judicious choice of the K_i can make the averaged eqs.(5.16) easier to handle. For example, if we take K_1=0 and

(5.17) $$K_3 = \frac{43\,\bar{a}^7 - 312\,\bar{a}^5 + 576\,\bar{a}^3}{-6144\,\bar{a}^2 + 8192}$$

then the averaged eqs.(5.16) become:

(5.18) $$\dot{\bar{a}} = \epsilon\,\frac{\bar{a}}{8}\,(4 - \bar{a}^2) + O(\epsilon^4), \quad \dot{\bar{\varphi}} = 1 - \epsilon^2\,\frac{11\,\bar{a}^4 - 48\,\bar{a}^2 + 32}{256} + O(\epsilon^4)$$

Note that in (5.18) the ϵ^2 term in the $\dot{\bar{a}}$ eq. has been eliminated by choosing $K_1=0$ and the ϵ^3 term has been eliminated by choosing K_3 as in eq.(5.17).

6. HOPF BIFURCATIONS

As an example of the application of computer algebra to bifurcation theory, we consider the following example [11]:

(6.1) $$\ddot{z} + z = \epsilon\left[\alpha_1\, z^2 + \alpha_2\, z\, \dot{z} + \alpha_3\, \dot{z}^2\right] + \epsilon^2\left[\mu\, \dot{z} + \beta_1\, z^3 + \beta_2\, z^2 \dot{z} + \beta_3\, z\, \dot{z}^2 + \beta_4\, \dot{z}^3\right]$$

where μ, α_i, β_i are parameters, and where $\epsilon \ll 1$. As in the case of van der Pol's equation considered in the previous section, eq.(6.1) may exhibit a periodic limit cycle solution for small ϵ. We shall be interested in understanding how such a periodic solution comes to be born as the parameters are varied.

We use second-order averaging on eq.(6.1). As in the previous section, we set

(6.2) $$z = a \cos\varphi\ , \quad \dot{z} = -\,a \sin\varphi$$

and apply a near-identity transformation as in the previous section, resulting in the averaged eqs.:

$$\dot{\bar{a}} = \epsilon^2 \left[\frac{\mu \bar{a}}{2} + \frac{1}{8} \left[\beta_2 + 3\,\beta_4 + \alpha_2(\alpha_1+\alpha_3)\right] \bar{a}^3 \right] + O(\epsilon^3) \tag{6.3}$$

and a similar eq. on $\dot{\bar{\varphi}}$ which we shall not be concerned with. A limit cycle solution to eq.(6.1) will correspond to a non-zero equilibrium solution to (6.3), i.e. to a real root $\bar{a}$ of the equation:

$$\frac{\mu}{2} + \frac{1}{8} \left[\beta_2 + 3\,\beta_4 + \alpha_2(\alpha_1+\alpha_3) \right] \bar{a}^2 = 0 \tag{6.4}$$

Thus a limit cycle is predicted to occur if $\beta_2+3\,\beta_4+\alpha_2(\alpha_1+\alpha_3)$ has the opposite sign of μ. If $\beta_2+3\,\beta_4+\alpha_2(\alpha_1+\alpha_3) \neq 0$, then a limit cycle will be born out of the origin $\bar{a} = 0$ as μ is swept through the value $\mu = 0$, a scenario which is known as a <u>Hopf bifurcation</u> [5].

7. AVERAGING WITH ELLIPTIC FUNCTIONS [10]

In the previous sections we applied regular perturbations and the method of averaging to d.e.'s which reduced to the simple harmonic oscillator when $\epsilon=0$, i.e. we perturbed off of sine and cosine solutions. In this section we shall apply the method of averaging to the equation

$$\ddot{x} + x + x^2 = \epsilon\, g(x,\dot{x}), \quad \epsilon \ll 1 \tag{7.1}$$

This involves perturbing off of the Jacobian elliptic function solutions of the equation

$$\ddot{x} + x + x^2 = 0 \tag{7.2}$$

This work illustrates the utility of computer algebra in facilitating the necessary computations.

We begin by briefly discussing Jacobian elliptic functions [2]. The functions cn and sn are the elliptic counterparts of the trigonometric functions cos and sin. Like cos, cn is even, while like sin, sn is odd. The identity

$$\mathrm{sn}^2 + \mathrm{cn}^2 = 1 \tag{7.3}$$

reminds us of the corresponding relation between sin and cos. The functions cn and sn are actually a family of functions which are parameterized by a "square-modulus" m:

$$\mathrm{cn} = \mathrm{cn}(u,m), \quad \mathrm{sn} = \mathrm{sn}(u,m) \tag{7.4}$$

In fact, cn and sn reduce to cos and sin for $m = 0$. The derivatives of cn and sn, however, introduce a new function, dn, which has no trigonometric counterpart:

$$\frac{d}{du}\,\mathrm{cn} = -\,\mathrm{sn}\;\mathrm{dn}, \quad \frac{d}{du}\,\mathrm{sn} = \mathrm{cn}\;\mathrm{dn} \tag{7.5}$$

In the trigonometric limit $m = 0$, dn reduces to unity. The function dn satisfies the following equations:

$$m\,\mathrm{sn}^2 + \mathrm{dn}^2 = 1, \quad \frac{d}{du}\,\mathrm{dn} = -\,m\;\mathrm{sn}\;\mathrm{cn} \tag{7.6}$$

The functions cn and sn are periodic with period $4K(m)$, where $K(m)$ is a tabulated function called the complete elliptic integral of the first kind. For m going from 0 to 1, $K(m)$ goes from $\pi/2$ to infinity. The function dn has period $2K(m)$.

A related quantity is E(m), the complete elliptic integral of the second kind. For m going from 0 to 1, E(m) goes from $\pi/2$ to 1. E(m) enters this work through the integral of cn^2 over one period 4K(m):

$$\int_0^{4K} cn^2 \, du = \frac{4}{m} \, [E(m) - (1-m) \, K(m)] \tag{7.7}$$

Next we obtain the solutions of eq.(7.2) which we are perturbing off of. Although the general solution to (7.2) may be found by utilizing the conservation of energy, evaluating the resulting elliptic integral and inverting, we find it more instructive to proceed as follows: Assume a solution in the form:

$$x = a_1 + a_2 \, sn^2, \qquad u = \omega \, t + b \tag{7.8}$$

where sn = sn(u,m) and where a_1, a_2, ω, b and m are constants. Differentiating (7.8) and using eq.(7.5) gives

$$\dot{x} = 2 \, a_2 \, \dot{u} \, sn \, sn' = 2 \, a_2 \, \omega \, sn \, cn \, dn \tag{7.9}$$

where ' represents differentiation with respect to u. Differentiating (7.9) and using eqs.(7.2)-(7.6) gives

$$\ddot{x} = 2 \, a_2 \, \omega^2 \, [(m-1) + 2(1-2m) \, cn^2 + 3m \, cn^4] \tag{7.10}$$

Substituting (7.10) into (7.2), collecting terms, and equating to zero the coefficients of cn^m, m=0,2,4, gives 3 eqs. for a_1, a_2 and ω, which may be solved as [10]:

(7.11) $$a_1 = \frac{1 + m - \sqrt{\lambda}}{2\sqrt{\lambda}}$$

(7.12) $$a_2 = -\frac{3m}{2\sqrt{\lambda}}$$

(7.13) $$\omega = \frac{1}{2\lambda^{1/4}}$$

where

(7.14) $$\lambda = m^2 - m + 1$$

Thus, the general solution, eq.(7.8), to eq.(7.2) involves the two arbitrary constants m and b. Here b determines the phase of the solution and m (where $0 \leq m \leq 1$) determines its amplitude.

In order to set up the method of averaging for eq.(7.1), we begin by using variation of parameters to express the effect of the order ϵ terms on the slow evolution of the square-modulus m, now considered a function of t.

We look for a solution to eq.(7.1) in the form of eq.(7.8), in which the two arbitrary constants m and b are allowed to vary in time. This results in first-order differential equations on m(t) and b(t). For brevity, we shall consider only the equation on m(t), the equilibria of which will correspond to periodic motions in eq.(7.1) (limit cycles).

Let us write the general solution, eq.(7.8), in the abstract form:

(7.15) $$x = x(t,m,b)$$

Differentiating eq.(7.15) gives

(7.16) $$\frac{dx}{dt} = \frac{\partial x}{\partial t} + \frac{\partial x}{\partial m}\frac{dm}{dt} + \frac{\partial x}{\partial b}\frac{db}{dt}$$

As usual in variation of parameters, we require that $\dot{x}$ be given by eq.(7.9), i.e.,

(7.17) $$\frac{dx}{dt} = \frac{\partial x}{\partial t}$$

giving

(7.18) $$\frac{\partial x}{\partial m}\frac{dm}{dt} + \frac{\partial x}{\partial b}\frac{db}{dt} = 0$$

Differentiating eq.(7.17) gives

(7.19) $$\frac{d^2x}{dt^2} = \frac{\partial^2 x}{\partial t^2} + \frac{\partial^2 x}{\partial m \partial t}\frac{dm}{dt} + \frac{\partial^2 x}{\partial b \partial t}\frac{db}{dt}$$

in which the unperturbed solution satisfies eq.(7.2), now written as

(7.20) $$\frac{\partial^2 x}{\partial t^2} + x + x^2 = 0$$

since m and b are held constant in eq.(7.20). In eq.(7.1), however, m and b are now permitted to depend on t, so that

(7.21) $$\frac{d^2x}{dt^2} + x + x^2 = \epsilon\ g$$

Substituting Eqs.(7.20) and (7.21) into Eq.(7.19) gives

(7.22) $$\frac{\partial^2 x}{\partial m \partial t}\frac{dm}{dt} + \frac{\partial^2 x}{\partial b \partial t}\frac{db}{dt} = \epsilon\ g$$

Eqs.(7.18) and (7.22) may be solved simultaneously for $\dot{m}$ as follows:

$$\dot{m} = \frac{\epsilon\ g\ x_b}{x_{mt}x_b - x_{bt}x_m} \qquad (7.23)$$

where subscripts represent partial differentiation.

The following **MACSYMA** program computes the partial derivatives in eq.(7.23) and simplifies the result:

```
depends([cn,sn,dn],[u,m]);
depends(u,[t,w,b]);
depends([a1,a2,w],m);
derivs:['diff(cn,u)=-sn*dn,'diff(sn,u)=cn*dn,'diff(dn,u)=-m*sn*cn,
        'diff(cn,m)=sn*dn*f,'diff(sn,m)=-cn*dn*f,
        'diff(dn,m)=m*sn*cn*f-sn^2/(2*dn)];
idents:[sn=sqrt(1-cn^2), dn= sqrt(1-m+m*cn^2)];
x:a1+a2*sn^2;
xt:diff(x,t);
xt:xt,derivs;
xbt:diff(xt,b);
xbt:xbt,derivs;
xb:diff(x,b);
xb:xb,derivs;
xm:diff(x,m);
xm:xm,derivs;
xmt:diff(xt,m);
xmt:xmt,derivs;
dmdt:-xb*e*g/(xm*xbt-xmt*xb);
dmdt:ratsimp(dmdt);
dmdt:dmdt,idents,diff;
dmdt:dmdt,derivs;
dmdt:dmdt,u=w*t+b,diff;
dmdt:dmdt,
a1 = -(sqrt(m^2-m+1)-m-1)/(2*sqrt(m^2-m+1)),
a2 = -3*m/(2*sqrt(m^2-m+1)),w = 1/(2*(m^2-m+1)^(1/4)),diff;
dmdt:factor(ratsimp(dmdt));
```

In writing the above program, the following identities were used ([2], p.283, formulas 710):

$$\frac{\partial sn}{\partial m} = -\ cn\ dn\ f,\ \frac{\partial cn}{\partial m} = cn\ dn\ f,\ \frac{\partial dn}{\partial m} = m\ sn\ cn\ f - \frac{1}{2}\,\frac{sn^2}{dn}$$

where

$$f = f(u,m) = \frac{E(u)-(1-m)u}{2m(1-m)} - \frac{sn\ cn}{2(1-m)dn}$$

Here is a record of the final expression produced by the program:

```
                 2                   2              2           7/4
8 CN SQRT(1 - CN ) E G SQRT((CN  - 1) M + 1) (M  - M + 1)
----------------------------------------------------------------
                            3 (M - 1)
```

That is,

(7.24) $$\dot{m} = -\frac{8}{3}\frac{\lambda^{7/4}}{1-m}\ sn\ cn\ dn\ \epsilon\ g$$

where $g = g(x,\dot{x}) = g(a_1+a_2\ sn^2, 2a_2\omega\ sn\ cn\ dn)$.

The right hand side (rhs) of eq.(7.24) is periodic in t and thus in the appropriate form for averaging. For first-order averaging, we may replace the rhs of eq.(7.24) by its average value taken over one period of length 4K(m):

(7.25) $$\dot{m} = -\frac{8}{3}\frac{\lambda^{7/4}}{1-m}\ \epsilon\ \frac{1}{4K}\int_0^{4K} sn\ cn\ dn\ g\ du$$

where $g = g(a_1+a_2\ sn^2, 2a_2\omega\ sn\ cn\ dn)$ and sn = sn(u,m), cn = cn(u,m) and dn = dn(u,m).

If $g(x,\dot{x})$ is a polynomial in x and $\dot{x}$, then the evaluation of the integral in eq.(7.25) may be readily accomplished. Terms in g of the form $x^N \dot{x}^M$ for M even lead to integrals of the form

(7.26) $$\int_0^{4K} \mathrm{cn}^P \, \mathrm{sn} \, \mathrm{dn} \, du$$

where P is an integer, which vanish due to the oddness of the integrand. On the other hand, terms in g of the form $x^N \dot{x}^M$ for M odd lead to integrals of the form

(7.27) $$\int_0^{4K} \mathrm{cn}^{2P} \, du$$

which may be evaluated by using the following results from [2], pp.192–3, formulas 312: Define

(7.28) $$C_{2P} = \int_0^{4K} \mathrm{cn}^{2P} \, du$$

Then

(7.29) $$C_0 = 4K, \quad C_2 = \frac{4}{m} \, [E-(1-m)K]$$

and

(7.30) $$C_{2P+2} = \frac{2P}{2P+1} \frac{2m-1}{m} C_{2P} + \frac{2P-1}{2P+1} \frac{1-m}{m} C_{2P-2}$$

where $E = E(m)$ and $K = K(m)$ are complete elliptic integrals.

An equilibrium point of the averaged $\dot{m}$ equation, eq.(7.25), corresponds to a limit cycle in the original equation (1). Thus if $m = m_0$ is a root of the equation

(7.31) $$\int_0^{4K} \mathrm{sn} \, \mathrm{cn} \, \mathrm{dn} \, g \, du = 0$$

then the averaged equation predicts that for small ϵ, a limit cycle coincides with the solution (7.8) of eq.(7.2) associated with a value of $m = m_0$.

For an arbitrary polynomial $g(x,\dot{x})$, the limit cycle integral condition, eq.(7.31), may be efficiently evaluated by using the following **MACSYMA** program:

```
g:read("Enter g(x,y), where y=dx/dt");
/* eliminate even powers of y */
g1:ev(g,y=-y);
g2:(g-g1)/2;
/* form integrand of limit cycle integral, eq.(7.31) */
integrand:g2*sn*cn*dn,x=a1+a2*sn^2,y=2*a2*w*sn*cn*dn$
/* use identities to express sn,dn in terms of cn */
idents:[sn=sqrt(1-cn^2), dn= sqrt(1-m+m*cn^2)];
integrand1:expand(ev(integrand,idents))$
/* set up rules for evaluating the integral of cn^r */
c[0]:4*k(m);
c[2]:4*(e(m)-(1-m)*k(m))/m;
c[r]:=(r-2)*(2*m-1)/((r-1)*m)*c[r-2]+(r-3)*(1-m)/((r-1)*m)*c[r-4];
/* perform the integration by replacing cn^r by c[r] */
max:hipow(integrand1,cn);
for i:0 thru max step 2 do b[i]:coeff(integrand1,cn,i);
result:sum(b[2*i]*c[2*i],i,0,max/2)$
/* substitute the parameters a1,a2,w in terms of m */
result1:result,
        a1 = -(sqrt(m^2-m+1)-m-1)/(2*sqrt(m^2-m+1)),
        a2 = -3*m/(2*sqrt(m^2-m+1)), w = 1/(2*(m^2-m+1)^(1/4))$
/* simplify the result */
result2:ratsimp(result1);
```

As an example let us take

$$\ddot{x} + x + x^2 = \epsilon\ (k_{01}\ \dot{x} + k_{11}\ x\ \dot{x}) \tag{7.32}$$

in which k_{01} and k_{11} are given parameters.

Here is a partial record of a sample run in which only the first and last lines of the program are displayed:

```
(C3) g:read("Enter g(x,y), where y=dx/dt");

Enter g(x,y), where y=dx/dt

K01*Y+K11*X*Y;

(D3)                    K11 X Y + K01 Y
```

. . .

```
(C16) /* simplify the result */

result2:ratsimp(result1);

                  2                    2
(D16) - (SQRT(M  - M + 1) ((7 K11 M  - 21 K11 M + 14 K11) K(M)

                 2
 + (- 14 K11 M  + 14 K11 M - 14 K11) E(M))

          2                    2
 + SQRT(M  - M + 1) ((- 14 K01 M  + 42 K01 M - 28 K01) K(M)

               2
 + (28 K01 M  - 28 K01 M + 28 K01) E(M))

            3          2
 + (5 K11 M  + 5 K11 M  - 20 K11 M + 10 K11) K(M)

               3           2
 + (- 10 K11 M  + 15 K11 M  + 15 K11 M - 10 K11) E(M))

    2          1/4      3       2
/((M  - M + 1)    (35 M  - 35 M  + 35 M))
```

This condition may be expressed in a convenient form by solving for k_{01}:

```
(C18) SOLVE(RESULT2,K01);
```

```
                           3            2                    2
(D18) [K01 = ((5 K11 M  + SQRT(M  - M + 1) (7 K11 M  - 21 K11 M + 14 K11)

              2                                                      3
 + 5 K11 M  - 20 K11 M + 10 K11) K(M) + (- 10 K11 M

            2                 2                                          2
 + SQRT(M  - M + 1) (- 14 K11 M  + 14 K11 M - 14 K11) + 15 K11 M

                                                      2                   2
 + 15 K11 M - 10 K11) E(M))/(SQRT(M  - M + 1) (14 M  - 42 M + 28) K(M)

              2                         2
 + (- 28 M  + 28 M - 28) SQRT(M  - M + 1) E(M))]
```

This result can be rewritten in the form:

$$\frac{k_{01}}{k_{11}} = \frac{1}{2} + \frac{5}{14}\,\frac{(K-2E)m^3+(K+3E)m^2+(-4K+3E)m+2(K-E)}{\sqrt{\lambda}\;\left[(K-2E)m^2+(-3K+2E)m+2(K-E)\right]} \tag{7.33}$$

where λ is given by eq.(7.14) and where K and E are complete elliptic integrals.

Numerical evaluation of eq.(7.33) shows [10] that averaging predicts no limit cycle if $k_{01}/k_{11}>1/7$ or if $k_{01}/k_{11}<0$, and a single limit cycle for $0<k_{01}/k_{11}<1/7$. In order to check this result, we numerically integrated eq.(7.32) using a Runge-Kutta scheme and found excellent agreement with the averaging predictions.

8. TWO VARIABLE EXPANSION METHOD

The two variable expansion method [7,11] is a popular singular perturbation method which offers a convenient alternative to averaging. It is essentially equivalent to the method of multiple scales [7]. The method may be motivated by noting that the method of averaging produces approximate solutions of the form (cf.eqs.(5.2),(5.3)):

$$z = a \cos \varphi \tag{8.1}$$

where

$$\frac{da}{dt} = O(\epsilon) , \qquad \frac{d\varphi}{dt} = 1 + O(\epsilon) \tag{8.2}$$

That is, $a = a(\epsilon t)$ and $\varphi = t + \psi(\epsilon t) = \varphi(t,\epsilon t)$. Thus we are led to the key ansatz that $z = z(t,\epsilon t)$. In order to simplify notation, we will use ξ and η to represent ordinary time and slow time, respectively:

$$z = z(\xi,\eta), \qquad \xi = t, \quad \eta = \epsilon t \tag{8.3}$$

In order to offer a comparison with the method of averaging presented in section 5, we will use the two variable expansion method on the same class of problems, cf.eq.(5.1):

$$\frac{d^2z}{dt^2} + z = \epsilon \; F(z,\frac{dz}{dt}) \tag{8.4}$$

Substitution of eq.(8.3) into (8.4) turns the o.d.e.(8.4) into a p.d.e., since now both and ξ and η are treated as independent variables:

$$\frac{dz}{dt} = \frac{\partial z}{\partial \xi}\frac{d\xi}{dt} + \frac{\partial z}{\partial \eta}\frac{d\eta}{dt} = \frac{\partial z}{\partial \xi} + \epsilon \; \frac{\partial z}{\partial \eta} \tag{8.5}$$

(8.6) $$\frac{d^2z}{dt^2} = \frac{\partial^2 z}{\partial \xi^2} + 2\epsilon \frac{\partial^2 z}{\partial \xi \partial \eta} + \epsilon^2 \frac{\partial^2 z}{\partial \eta^2}$$

which gives

(8.7) $$\frac{\partial^2 z}{\partial \xi^2} + 2\epsilon \frac{\partial^2 z}{\partial \xi \partial \eta} + \epsilon^2 \frac{\partial^2 z}{\partial \eta^2} + z = \epsilon\ F(z, \frac{\partial z}{\partial \xi} + \epsilon \frac{\partial z}{\partial \eta})$$

Now we expand z in a power series in ϵ:

(8.8) $$z(\xi,\eta) = z_0(\xi,\eta) + \epsilon\ z_1(\xi,\eta) + 0(\epsilon^2)$$

Substituting eq.(8.8) into (8.7) and collecting terms gives:

(8.9) $$\frac{\partial^2 z_0}{\partial \xi^2} + z_0 = 0$$

(8.10) $$\frac{\partial^2 z_1}{\partial \xi^2} + z_1 = -\ 2 \frac{\partial^2 z_0}{\partial \xi \partial \eta} + F(z_0, \frac{\partial z_0}{\partial \xi})$$

The general solution to the p.d.e.(8.9) is:

(8.11) $$z_0 = a(\eta)\ \cos(\xi + \psi(\eta))$$

Note that since (8.9) is a p.d.e., the "constants of integration" a and ψ become arbitrary functions of the variable η.

Substitution of eq.(8.11) into (8.10) gives a p.d.e. on $z_1(\xi,\eta)$:

(8.12) $$\frac{\partial^2 z_1}{\partial \xi^2} + z_1 = 2 \frac{\partial a}{\partial \eta} \sin(\xi+\psi) - 2a \frac{\partial \psi}{\partial \eta} \cos(\xi+\psi)$$
$$+ F(a \cos(\xi+\psi), - a \sin(\xi+\psi))$$

For a particular choice of the function $F(z,\dot{z})$, the right-hand side of eq.(8.12) may be trigonometrically reduced, and terms may be collected as follows:

$$(8.13)\qquad \frac{\partial^2 z_1}{\partial \xi^2} + z_1 = (\ldots)\ \sin(\xi+\psi) + (\ldots)\ \cos(\xi+\psi) + \text{higher harmonics}$$

In order to avoid "secular" terms, i.e., terms which grow linearly in ξ and which prevent the resulting approximate solution from being uniformly valid in the limit as t approaches infinity, we require the coefficients $(\ldots)$ of $\sin(\xi+\psi)$ and $\cos(\xi+\psi)$ to vanish. This gives two o.d.e.'s on $a(\eta)$ and $\psi(\eta)$, which, when solved and substituted into eq.(8.11), provide the general solution to the original problem (8.4), accurate to $O(\epsilon)$.

As an example, we again take van der Pol's eq.(5.6):

$$(8.14)\qquad \frac{d^2 z}{dt^2} + z = \epsilon\ \frac{dz}{dt}\ (1 - z^2)$$

Here the function $F(z,\dot{z})$ of (8.4) becomes

$$(8.15)\qquad F(z,\dot{z}) = \frac{dz}{dt}\ (1 - z^2)$$

F appears on the right-hand side of eq.(8.12) in the form:

$$(8.16)\ F(a \cos(\xi+\psi), - a \sin(\xi+\psi)) = - a \sin(\xi+\psi)\ (1 - a^2 \cos^2(\xi+\psi))$$

Trigonometric reduction of (8.16) gives:

$$(8.17)\qquad F(a \cos(\xi+\psi), - a \sin(\xi+\psi)) = (- a + \frac{a^3}{4})\ \sin(\xi+\psi) + \frac{a^3}{4} \sin 3(\xi+\psi)$$

Elimination of secular terms in eq.(8.12) gives the two o.d.e.'s:

$$(8.18) \qquad \sin(\xi+\psi): \; 2\,\frac{\partial a}{\partial\eta} = a - \frac{a^3}{4}\,, \qquad \cos(\xi+\psi): \; 2a\,\frac{\partial\psi}{\partial\eta} = 0$$

Note that these eqs. are equivalent to eqs.(5.15) obtained by the method of averaging. Since η represents slow time, eqs.(8.18) are called a slow-flow. Computer algebra may be used to solve the slow-flow on $a(\eta)$:

```
(C2) depends(a,eta);
(D2)                              [A(ETA)]
(C3) 2*diff(a,eta) = a-a^3/4;
                                            3
                                dA         A
(D3)                          2 ---- = A - --
                                dETA       4
(C4) ode2(%,a,eta);
(D4)     - LOG(A + 2) + 2 LOG(A) - LOG(A - 2) = ETA + %C
(C5) soln1:logcontract(%);
                                 2
                                A
(D5)                      LOG(------) = ETA + %C
                               2
                              A  - 4
(C6) %e^lhs(%) = %e^rhs(%);
                              2
                             A         ETA + %C
(D6)                       ------ = %E
                            2
                           A  - 4
(C7) soln2:solve(%,a^2);
                                    ETA + %C
                          2     4 %E
(D7)                    [A  = ---------------]
                                 ETA + %C
                              %E         - 1
```

```
     /* set initial condition */
(C8) soln1,a = a0,eta = 0;
                                     2
                                   A0
(D8)                          LOG(--------) = %C
                                    2
                                  A0  - 4
(C9) solve(%,%C);
                                         2
                                       A0
(D9)                      [%C = LOG(--------)]
                                        2
                                      A0  - 4
(C10) ratsimp(ev(soln2,%));

                                     2  ETA
                       2         4 A0  %E
(D10)                [A  = - --------------------]
                                2  ETA      2
                              A0 %E    - A0  + 4
```

That is,

$$a^2(\eta) = \frac{4\ a^2(0)\ e^{\eta}}{a^2(0)\ e^{\eta} - a^2(0) + 4} \tag{8.19}$$

This result, when substituted into eq.(8.11), gives the approximate solution:

$$z(t) = \frac{2}{\sqrt{1 + \left[\frac{4}{a^2(0)} - 1\right] e^{-\epsilon t}}} \cos(t + \psi(0)) + O(\epsilon) \tag{8.20}$$

Note that this solution contains two arbitrary constants $a(0)$, $\psi(0)$, and thus represents an approximate general solution to van der Pol's eq.(8.14). Note also that as $t \to +\infty$, $z \to 2\cos(t+\psi(0))$. For large t, binomial expansion of eq.(8.20) shows that the approach to the limit cycle goes like $e^{-\epsilon t}$. In the limit as $t \to -\infty$, there are two possibilities. If $a(0) < 2$, then $z \to 0$ as $t \to -\infty$. However, if $a(0) > 2$, then z blows up in finite time! This behavior should not surprise us, and is due to the polynomial nature of the nonlinearity in eq.(8.14). For example, the simple model equation

$$\frac{dz}{dt} = z^2 \tag{8.21}$$

has the general solution

$$z(t) = \frac{1}{\frac{1}{z(0)} - t} \tag{8.22}$$

which blows up in finite time $t = 1/z(0)$. Thus the approximate solution to van der Pol's eq., as generated by the two variable expansion method, captures numerous features of the behavior of the real solution. Numerical integration of eq.(8.14) shows good agreement with (8.20) for small ϵ.

The foregoing procedure may be formalized into a computer algebra program:

```
/* two variable expansion on z''+z=e f(z,z') */
depends([a,psi],eta);
z0:a*cos(xi+psi);
/* define f=f(z,zdot) here: */
f(z,zdot):=zdot*(1-z^2);
/* rhs= right hand side of z1 eq. */
rhs:-2*diff(z0,xi,1,eta,1)+f(z0,diff(z0,xi));
rhs2:expand(trigreduce(expand(%)));
eq1:coeff(rhs2,sin(xi+psi));
eq2:coeff(rhs2,cos(xi+psi));
solve([eq1,eq2],[diff(a,eta),diff(psi,eta)]);
```

Here is a sample run:

```
(C3) /* two variable expansion on z''+z=e f(z,z') */

depends([a,psi],eta);
(D3)                     [A(ETA), PSI(ETA)]
```

```
(C4) z0:a*cos(xi+psi);
(D4)                      A COS(XI + PSI)

(C5) /* define f=f(z,zdot) here: */

f(z,zdot):=zdot*(1-z^2);
                                              2
(D5)                F(Z, ZDOT) := ZDOT (1 - Z )

(C6) /* rhs= right hand side of z1 eq. */

rhs:-2*diff(z0,xi,1,eta,1)+f(z0,diff(z0,xi));

              dA                            dPSI
(D6) - 2 (- ---- SIN(XI + PSI) - A ---- COS(XI + PSI))
             dETA                           dETA

                                        2     2
                         - A (1 - A  COS  (XI + PSI)) SIN(XI + PSI)

(C7) rhs2:expand(trigreduce(expand(%)));
      3
     A  SIN(3 XI + 3 PSI)        dA
(D7) -------------------- + 2 ---- SIN(XI + PSI)
              4                 dETA

      3
     A  SIN(XI + PSI)                              dPSI
   + ----------------- - A SIN(XI + PSI) + 2 A ---- COS(XI + PSI)
              4                                   dETA

(C8) eq1:coeff(rhs2,sin(xi+psi));
                                      3
                                dA   A
(D8)                      2 ---- + -- - A
                               dETA   4

(C9) eq2:coeff(rhs2,cos(xi+psi));
                                  dPSI
(D9)                          2 A ----
                                  dETA

(C10) solve([eq1,eq2],[diff(a,eta),diff(psi,eta)]);

                                3
                      dA       A  - 4 A   dPSI
(D10)              [[---- = - ---------, ---- = 0]]
                     dETA          8      dETA
```

These results agree with eqs.(8.18).

Next we apply the foregoing program to the following example:

$$\frac{d^2z}{dt^2} + z = \epsilon \left[b_1 \frac{dz}{dt} + b_2 \left[\frac{dz}{dt}\right]^3 + b_3 \left[\frac{dz}{dt}\right]^5 \right] \tag{8.23}$$

where the b_i are given parameters. This entails replacing the line in the program which defines f, with the line:

(8.24)

```
f(z,zdot):=b1*zdot+b2*zdot^3+b3*zdot^5;
```

The resulting program gives the following slow-flow:

```
                 5         3
(D10)   [[dA   5 A  B3 + 6 A  B2 + 8 A B1   dPSI
          ---- = ------------------------, ---- = 0]]
          dETA             16               dETA
```

Limit cycles correspond to (non-zero) equilibria in the $dA/d\eta$ equation:

$$8\ b_1 + 6\ b_2\ \rho + 5\ b_3\ \rho^2 = 0, \qquad \rho = a^2 \tag{8.25}$$

A positive root ρ of (8.25) corresponds to a limit cycle. There may be 0, 1 or 2 limit cycles. For example, for $b_1 = 1$, $b_2 = -3$ and $b_3 = 1$, eq.(8.25) predicts the existence of two limit cycles with amplitudes a = 0.72 and 1.75, in agreement with numerical integration of eq.(8.23) for small ϵ.

9. APPLICATION TO COUPLED OSCILLATORS

In this section we apply the two variable expansion method to a system of n coupled oscillators of the form:

$$\frac{d^2 z_i}{dt^2} + z_i = \epsilon\, F_i(z_j, \frac{dz_j}{dt}) \,, \qquad i,j=1,2,\ldots,n \tag{9.1}$$

Note that the n oscillators are identical for $\epsilon = 0$. As in section 8, we begin by asserting that

$$z_i = z_i(\xi,\eta), \qquad \xi = t, \quad \eta = \epsilon t \tag{9.2}$$

Then eq.(9.1) becomes, neglecting terms of $O(\epsilon^2)$:

$$\frac{\partial^2 z_i}{\partial \xi^2} + 2\epsilon \frac{\partial^2 z_i}{\partial \xi \partial \eta} + z_i = \epsilon\, F_i(z_j, \frac{\partial z_j}{\partial \xi}) \tag{9.3}$$

We expand z_i in a power series in ϵ:

$$z_i(\xi,\eta) = z_{i0}(\xi,\eta) + \epsilon\, z_{i1}(\xi,\eta) + O(\epsilon^2) \tag{9.4}$$

which gives

$$\frac{\partial^2 z_{i0}}{\partial \xi^2} + z_{i0} = 0 \tag{9.5}$$

$$\frac{\partial^2 z_{i1}}{\partial \xi^2} + z_{i1} = -\, 2 \frac{\partial^2 z_{i0}}{\partial \xi \partial \eta} + F_i(z_{j0}, \frac{\partial z_{j0}}{\partial \xi}) \tag{9.6}$$

The general solution of eq.(9.5) is

$$z_{i0} = a_i(\eta) \cos(\xi + \psi_i(\eta)) \tag{9.7}$$

Substitution of (9.7) into (9.6) for a particular choice of the F_i permits terms to be arranged as follows:

$$\frac{\partial^2 z_{i1}}{\partial \xi^2} + z_{i1} = (\ldots) \sin \xi + (\ldots) \cos \xi + \text{higher harmonics} \tag{9.8}$$

Elimination of the secular terms (...) in each of the n eqs.(9.8) results in 2n first order o.d.e.'s governing the 2n unknowns

$$a_1, \psi_1, a_2, \psi_2, \ldots, a_n, \psi_n.$$

As an example, we consider a <u>ring</u> of n van der Pol oscillators with "diffusion" coupling:

$$\frac{d^2 z_i}{dt^2} + z_i = \epsilon \frac{dz_i}{dt} (1 - z_i^2) + \epsilon \alpha (z_{i+1} - 2 z_i + z_{i-1}) \tag{9.9}$$

where i=1,...,n, and where we use the convention $z_0 \equiv z_n$ and $z_{n+1} \equiv z_1$, because the oscillators are arranged in a ring.

The following **MACSYMA** program sets up the right-hand side of eq.(9.6), substitutes eq.(9.7) and collects secular terms as in eq.(9.8):

```
/* two variable expansion on a system of coupled oscillators */
/* n=number of oscillators */
n:read("enter n");
depends([a,psi],eta);
for i:1 thru n do
  z0[i]:a[i]*cos(xi+psi[i]);
/* define f[i] here: */
f[1]:diff(z0[1],xi)*(1-z0[1]^2)+alpha*(z0[2]-2*z0[1]+z0[n]);
for i:2 thru n-1 do
  f[i]:diff(z0[i],xi)*(1-z0[i]^2)
       +alpha*(z0[i+1]-2*z0[i]+z0[i-1]);
f[n]:diff(z0[n],xi)*(1-z0[n]^2)+alpha*(z0[1]-2*z0[n]+z0[n-1]);
/* rhs[i]= right hand side of z1[i] eq. */
for i:1 thru n do(
  rhs[i]:-2*diff(z0[i],xi,1,eta,1)+f[i],
  rhs2:expand(trigexpand(rhs[i])),
  rhs3:expand(trigreduce(rhs2,xi)),
  eq1:coeff(rhs3,sin(xi)),
  eq2:coeff(rhs3,cos(xi)),
  eqs[i]:solve([eq1,eq2],[diff(a[i],eta),diff(psi[i],eta)]));
slowflow:makelist(eqs[i],i,1,n)$
trigident:makelist(sin(psi[i])^2=1-cos(psi[i])^2,i,1,n)$
slowflow2:slowflow,trigident$
slowflow3:ratsimp(slowflow2)$
slowflow4:expand(trigreduce(expand(slowflow3)));
```

Although the computation can only be accomplished for a specific value of n, we can, by comparing the resulting slow-flows for n=3,4,5, obtain the following slow-flow which is valid for a general value of n:

$$\frac{da_i}{d\eta} = \frac{a_i}{2} - \frac{a_i^3}{8} + \frac{\alpha}{2}\left[a_{i+1}\sin(\psi_{i+1}-\psi_i) + a_{i-1}\sin(\psi_{i-1}-\psi_i)\right] \tag{9.10}$$

$$\frac{d\psi_i}{d\eta} = \alpha - \frac{\alpha}{2a_i}\left[a_{i+1}\cos(\psi_{i+1}-\psi_i) + a_{i-1}\cos(\psi_{i-1}-\psi_i)\right] \tag{9.11}$$

Eqs.(9.10) and (9.11) are valid for i=1,2,...,n with the convention that $a_0 \equiv a_n$, $\psi_0 \equiv \psi_n$, $a_{n+1} \equiv a_1$, $\psi_{n+1} \equiv \psi_1$. The complete analysis of these 2n slow-flow equations has not, to my knowledge, been accomplished. Here we discuss two simple solutions corresponding to equilibria of the slow-flow, i.e., to periodic motions of the original system (9.9).

In the case that $\alpha = 0$, i.e., no coupling, each oscillator exhibits, to $O(\epsilon)$, an amplitude of 2. We seek solutions such that when $\alpha \neq 0$ this behavior persists:

$$a_1 = a_2 = \cdots = a_n = 2 \tag{9.12}$$

Substitution of (9.12) into the slow-flow (9.10),(9.11) gives:

$$0 = \alpha \left[\sin(\psi_{i+1}-\psi_i) + \sin(\psi_{i-1}-\psi_i)\right] \tag{9.13}$$

$$\frac{d\psi_i}{d\eta} = \alpha - \frac{\alpha}{2} \left[\cos(\psi_{i+1}-\psi_i) + \cos(\psi_{i-1}-\psi_i)\right] \tag{9.14}$$

One solution to eqs.(9.13),(9.14) is

$$\psi_1 = \psi_2 = \cdots = \psi_n \tag{9.15}$$

whereupon

$$\frac{d\psi_i}{d\eta} = 0 \tag{9.16}$$

This motion corresponds to each of the oscillators vibrating in phase with the others. Their common frequency is, to $O(\epsilon)$, unity, cf.(9.7).

Another solution to eqs.(9.13),(9.14) is

$$\psi_{i+1} = \psi_i + \frac{2\pi}{n}, \qquad i = 1,2,\ldots,n-1 \tag{9.17}$$

or, equivalently,

$$\psi_i = \psi_1 + \frac{2\pi}{n}(i-1), \qquad i = 2,3,\ldots,n \tag{9.18}$$

Note that the slow-flow (9.10),(9.11) is 2π-periodic in ψ_i, so that all angles ψ_i are understood mod 2π. Thus there is no jump in (9.18) for i=n, i.e., $\psi_n = \psi_1 + \frac{2\pi}{n}(n-1)$ means $\psi_1 = \psi_n + \frac{2\pi}{n} \pmod{2\pi}$.

Substituting eq.(9.17) into (9.14) gives

$$\frac{d\psi_i}{d\eta} = \alpha \left[1 - \cos\frac{2\pi}{n}\right] \tag{9.19}$$

This motion corresponds to a wave moving around the ring. The phase of each oscillator differs from that of its neighbors by a constant amount, and each has a common frequency equal to

$$1 + \epsilon\, \alpha \left[1 - \cos\frac{2\pi}{n}\right] \tag{9.20}$$

For large n, this frequency may be approximated by $1 + \frac{2\pi^2 \epsilon \alpha}{n^2}$.

The two solutions (9.15) and (9.17) may be viewed as special cases of a more general class of solutions to (9.13),(9.14). Consider the solution:

$$\psi_{i+1} = \psi_i + \frac{2m\pi}{n}, \qquad i = 1,2,\ldots,n-1 \tag{9.21}$$

where $m < n$ is an integer. This solution may also be viewed as a wave moving around the ring, but now with m complete wavelengths exhibited in one ring circumference. From this point of view, the uniform solution (9.15) corresponds to m=0 and the solution (9.17) corresponds to m=1.

A related question concerns the stability and bifurcation of all of these motions. Although we will not consider this here, we note that this computation is more conveniently accomplished via the slow-flow (9.13),(9.14), where the motion in question is an equilibrium point, than via the original eqs.(9.9), where the motion is a function of time.

REFERENCES

1. C.M.Bender and S.A.Orszag, "Advanced Mathematical Methods for Scientists and Engineers", McGraw-Hill (1978)

2. P.Byrd and M.Friedman, "Handbook of Elliptic Integrals for Engineers and Physicists", Springer (1954)

3. V.T.Coppola and R.H.Rand,"Averaging Using Elliptic Functions: Approximation of Limit Cycles", Acta Mechanica, 81:125-142 (1990)

4 V.T.Coppola and R.H.Rand, "MACSYMA Program to Implement Averaging Using Elliptic Functions", in Computer Aided Proofs in Analysis, eds. K.R.Meyer and D.S.Schmidt, pp.71–89, Springer (1991)

5. J.Guckenheimer and P.Holmes, "Nonlinear Oscillations, Dynamical Systems and Bifurcations of Vector Fields", Springer (1983)

6. P.B.Kahn and Y.Zarmi, "Minimal Normal Forms in Harmonic Oscillations with Small Nonlinear Perturbations", Physica D, 54:65-74 (1991)

7. A.Nayfeh, "Perturbation Methods", Wiley (1973)

8. R.H.Rand, "Computer Algebra in Applied Mathematics: An Introduction to MACSYMA", Pitman (1984)

9. R.H.Rand, "The Use of Symbolic Computation in Perturbation Analysis", in Symbolic Computation in Fluid Mechanics and Heat Transfer, ed. H.H.Bau et al., ASME publication HTD-105/AMD-97, pp.41-45 (1988)

10. R.H.Rand, "Using Computer Algebra to Handle Elliptic Functions in the Method of Averaging", in Symbolic Computations and Their Impact on Mechanics, eds. A.K.Noor,I.Elishakoff, G.Hulbert, pp.311–326, Amer.Soc.Mech.Eng., PVP-Vol.205, (1990)

11. R.H.Rand and D.Armbruster, "Perturbation Methods, Bifurcation Theory and Computer Algebra", Springer (1987)

12. J.A.Sanders and F.Verhulst, "Averaging Methods in Nonlinear Dynamical Systems", Springer (1985)

13. J.J.Stoker, "Nonlinear Vibrations", Interscience (1950)

14. M.Van Dyke, "Analysis and Improvement of Perturbation Series", Q.J.Mech.Appl.Math., 27:423–450 (1974)

ANALYSIS OF NONLINEAR DYNAMIC ENGINEERING SYSTEMS

W. Kleczka
Technical University Hamburg-Harburg, Hamburg, Germany

Abstract

An integrated concept with numerical and symbolical parts for a detailed investigation of periodic motions of nonlinear mechanical systems described by ordinary differential equations is presented. Thereby, the local stability analysis is based upon a local approximation of the Poincaré map which has to be derived in the neighborhood of the periodic solution to be investigated. If a bifurcation takes place the system can be reduced to a critical subsystem containing all information about the local nonlinear behavior. Afterwards, the system is transformed to a nonlinear normal form which allows for a classification of the bifurcation scenario.

Beginning with the derivation of the mathematical model, the entire procedure is applied for the stability analysis of a periodically forced submerged double pendulum. Thereby, special difficulties due to the non-smooth characteristics of the damping forces have to be overcome.

The application of computer algebra software is prerequisite for the presented approach because the analytical operations and expressions become rather involved. It is of major importance that the developed computer tools are not specific to a problem.

Symbolic computation stretches the limits of analyzability far beyond a simple pen-and-paper approach. Computer power, however, is still the limiting factor because especially for high-dimensional systems computer algebra software can already operate at its limits with respect to memory requirements and computation time.

The combination of numerical and symbolical procedures is a very promising approach for a highly sophisticated analysis of nonlinear engineering problems. By means of computer-aided tools a systematic analysis of bifurcation phenomena is visible.

1 Introduction

The increasing level of computer power together with the availability of high performance scientific software undoubtedly allows more adequate models of technical systems in the sense that these models are more accurate but still derivable, analyzable, and solvable in a reasonable amount of time and with sufficient reliability.

Consequently, almost all engineering problems, especially in mechanics, result in nonlinear mathematical models. The performance of a technical system depends on its dynamic behavior which can be predicted using these mathematical models. Changes of system parameters may cause changes in the qualitative behavior, so-called bifurcations. This stability problem is particularly important in the study of periodic oscillations of nonlinear systems.

1.1 Overview

The traditional pen-and-paper approach for the derivation and analysis of nonlinear mathematical models of technical systems is tedious and error-prone. The only way to come to a systematic analysis of nonlinear dynamical engineering systems was and still is the development of computer supported tools. The first tools applicable not only to a specific system but to a class of systems have been purely numerical, e.g. KREUZER (1987). The following methods have proven to be generally applicable and useful:

- Cell mapping approach, e.g. HSU (1980).
- Calculation of Lyapunov exponents, e.g. KLECZKA (1985).
- Calculation of fixed points and periodic solutions, e.g. WILMERS (1988).

These tools are not designated to provide an overall insight into the system under consideration even if the experienced expert can draw conclusions from such kind of numerical results. With these tools the engineer is able to speed up the analysis as the time consuming generation and critical examination of endless trajectories can be omitted. On the background of the exponential development of computing power and forthcoming parallel computers these numerical tools promise to be very powerful in the near future even for higher-dimensional practical engineering problems.

On the other hand, computer support is desirable not only for numerical purposes. Analytical considerations provide a deeper insight into the system behavior and allow for a broader understanding of the sometimes very complex phenomena especially in the field of nonlinear dynamics.

With the advent of appropriate software computerized symbol manipulation became possible and gave an important impact for the development of computer-aided analytical tools. Consequently, the next step for the systematic analysis of nonlinear dynamic systems was the application of analytical concepts, e.g. by means of computer algebra systems such as Maple, CHAR ET AL. (1991). In this domain appreciable approaches are:

- Center manifold formalism, e.g. RAND AND ARMBRUSTER (1987).
- Local approximation of the Poincaré map, e.g. KLECZKA (1989).
- Normal form transformation, e.g. LICHTENBERG (1992).

These methods are relevant and applicable for a wide range of nonlinear dynamical systems. They allow a detailed investigation that affords an insight into the dynamical structure of the system. Only a systematic and widely automated approach together with the enormous power of today's computers make a reliable application of these methods to more complicated engineering systems possible.

From an engineering point of view it is important to have some kind of toolbox for the analysis of nonlinear systems in parameter space. In combination with reliable standard numerical methods a systematic computer-aided concept for the local stability analysis of nonlinear dynamic systems with periodic excitation could be developed.

1.2 Problem Formulation

Dynamical systems governed by a set of ordinary differential equations

$$\dot{\boldsymbol{x}} = \boldsymbol{f}(\boldsymbol{x}, \boldsymbol{\mu}, t) \tag{1.1}$$

are considered. Here, t denotes the time, the state vector is given by $\boldsymbol{x} \in \mathbb{R}^n$, the right hand sides by the vector-valued function $\boldsymbol{f} \in \mathbb{R}^n$, and the parameter vector by $\boldsymbol{\mu} \in \mathbb{R}^{n_r}$. The question is: how does the equilibrium or periodic orbit of a vector field described by (1.1) change with parameter perturbations? Bifurcation theory provides methods to examine this problem which are usually based on multivariable calculus and differential topology.

Instead of looking at trajectories in the time domain the so-called *Poincaré map*

$$\boldsymbol{x}_{N+1} = \boldsymbol{p}(\boldsymbol{x}_N, \boldsymbol{\mu}) \tag{1.2}$$

is used for studying the stability of periodic solutions $\bar{\boldsymbol{x}}(t) = \bar{\boldsymbol{x}}(t+T)$ with period T. Again, $\boldsymbol{p}$ is a nonlinear vector-valued function which is smooth; the discrete time is denoted by N. If the discretization time of the point map (1.2) is chosen to be the period T, then the respective periodic solution of the continuous system (1.1) corresponds to a fixed point of the discrete system (1.2).

1.3 Computer-Aided Analysis Concept for Local Bifurcations

The logical structure of the overall-analysis is outlined in Fig. 1.1. Usually the first step is to map the considered real-world physical system onto a simplified mechanical system and to determine relevant parameter ranges. The derivation of a mathematical model is the next step. For engineering problems represented by multibody systems this task can be symbolically performed by the special-purpose symbol manipulation package Neweul, KREUZER AND SCHIEHLEN (1990), KREUZER AND LEISTER (1991). It follows the detection of periodic solutions by numerical routines. In order to analyze the stability of the motion a local approximation of the Poincaré Map is generated by a combined analytic-numeric procedure which results in numerical coefficients of a truncated *Taylor series* expansion, KLECZKA (1989). If the nonlinear system is uncritical the stability analysis can be restricted to the linearized map; this leads to the well-known *Jordan normal form* for linear systems. By means of an iterative procedure an approximation for the bifurcation value of the control parameter can be calculated on the basis of the linearized map with nonlinear parameter dependence. Linear stability analysis fails for nonlinear systems with critical eigenvalues, i.e. eigenvalues for which a bifurcation takes place. It can be shown that for this case after decoupling the linear part of the semi-critical model the system can be reduced to a lower-dimensional critical subsystem without losing information about the stability properties, CARR (1981). Afterwards, the reduced system is transformed to a *nonlinear normal form*. Once the normal form has been computed one can investigate the dynamics of the normal form and classify the situation under consideration. This reduction and transformation process is performed by specifically designed routines implemented in Maple which proved to be a very reliable computer algebra system for this type of problems, KLECZKA ET AL. (1991).

The different analysis steps are enumerated and explained below (S stands for symbolic computations, N for numeric computations, and S/N for coupled symbolic-numeric computations):

1. S Generation of the equations of motion via Neweul.
2. N Detection of periodic solutions.
3. S/N Local approximation of the Poincaré map.
4. S/N Detection of bifurcation points.
5. S/N Linear decoupling.
6. S Center manifold reduction.
7. S Normal form transformation.

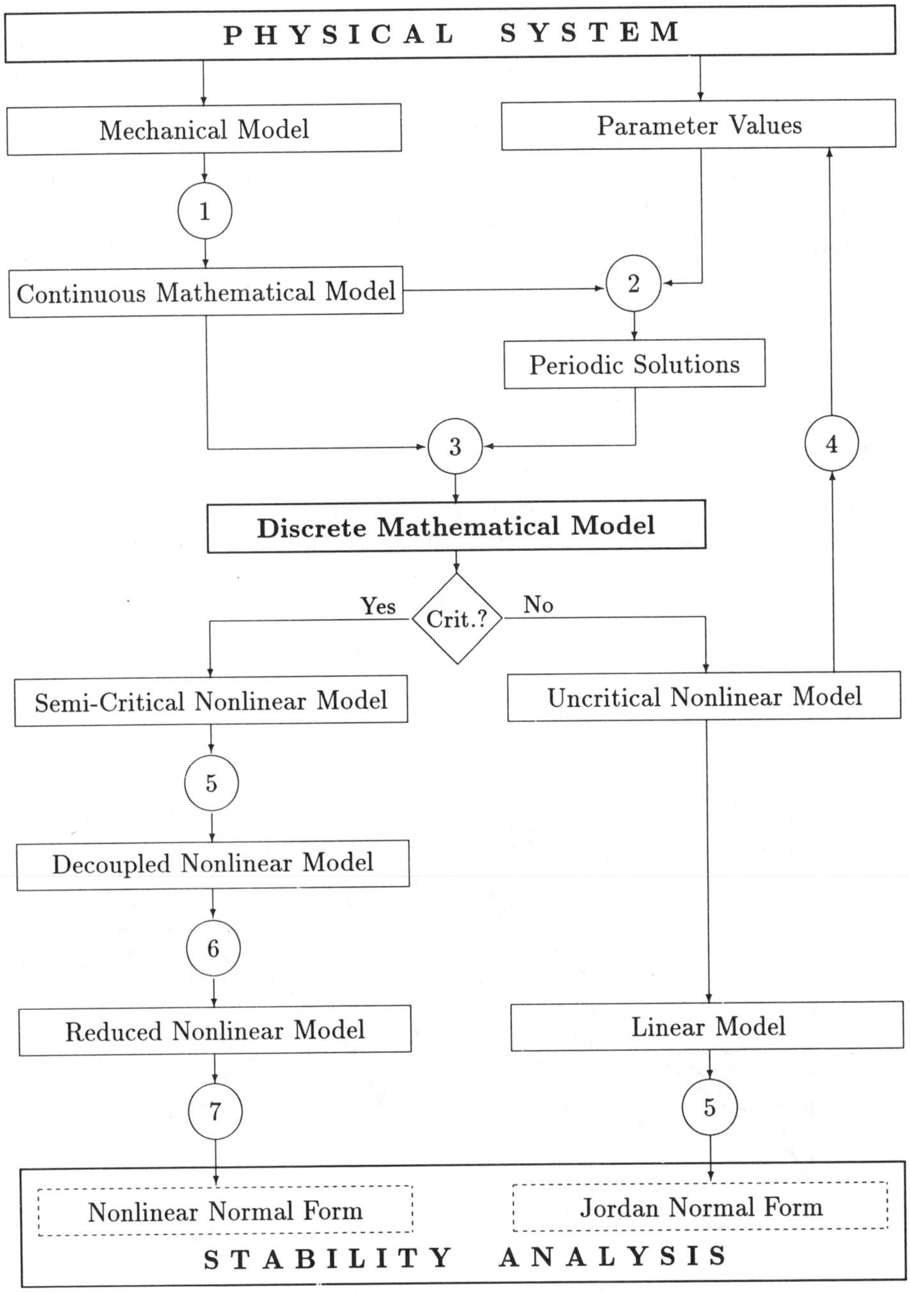

Figure 1.1: Logical structure of the analysis

2 Poincaré Maps

The discretization of time by using a Poincaré surface of section is a well-known and very useful method for the simplification of the analysis of a continuous dynamical system. Unfortunately, in general the point mapping cannot be derived in an analytical form; the global Poincaré map can be determined only by numerical integration of the underlying continuous system. A method is presented how to derive a local symbolic approximation of the Poincaré map in the neighborhood of the periodic solution to be studied. For the computer algebra implementation this method is strongly optimized in order not to be restricted to trivial problems.

2.1 Fundamentals

A Poincaré surface is locally defined as an $(n-1)$–dimensional hyper-plane Σ in the n–dimensional phase space which cuts the flow *transversely*. The transversality condition can also be satisfied for curved surfaces as can be seen in Fig. 2.1.

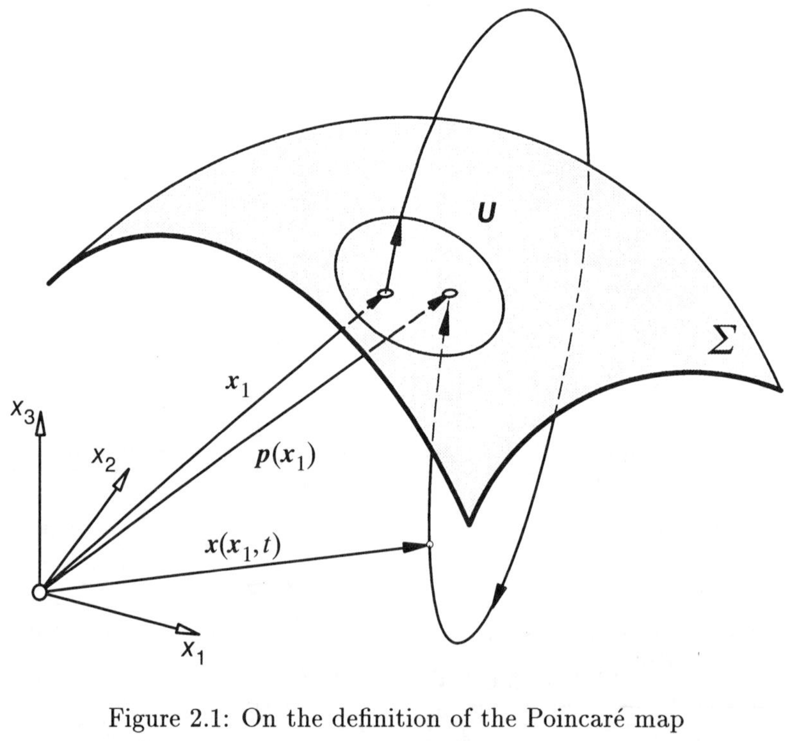

Figure 2.1: On the definition of the Poincaré map

Describing the local dynamics of the system in terms of the sequence of piercing points of the flow ϕ through this surface leads to the Poincaré map

$$\boldsymbol{p}(\boldsymbol{x}_1) = \boldsymbol{\phi}(\boldsymbol{x}_1, T_0) \tag{2.1}$$

where $T_0 = T_0(\boldsymbol{x}_1)$ is the time which the orbit $\boldsymbol{\phi}(\boldsymbol{x}_1)$ — starting at $\boldsymbol{x}_1$ — needs to reach Σ again.

Thereby, the dimension of the state space is reduced by one; instead of a time-continuous differential equation a 'simple' point mapping has to be considered:

$$\dot{\boldsymbol{x}} = \boldsymbol{f}(\boldsymbol{x}) \quad \longrightarrow \quad \boldsymbol{x}_{t+T_0} = \boldsymbol{p}(\boldsymbol{x}_t) \; . \tag{2.2}$$

For a periodically excited system the Poincaré surface can be globally defined because of the special geometry of the phase space:

$$\Sigma = \left\{ (\boldsymbol{x}, \theta) \in \mathbb{R}^n \times \mathsf{S}^1 \mid \theta = \theta_0 \right\} \; , \tag{2.3}$$

where the unit circle $\mathsf{S}^1 = \mathbb{R}$ (mod T) represents the periodicity $T = \frac{2\pi}{\omega}$ of the vector field $\boldsymbol{f}$. If the phase angle θ_0 is fixed, which determines the position of the surface of section relative to the phase angle of the excitation, then by Eq. (2.1) the Poincaré map of this surface Σ onto itself is given, Fig. 2.2.

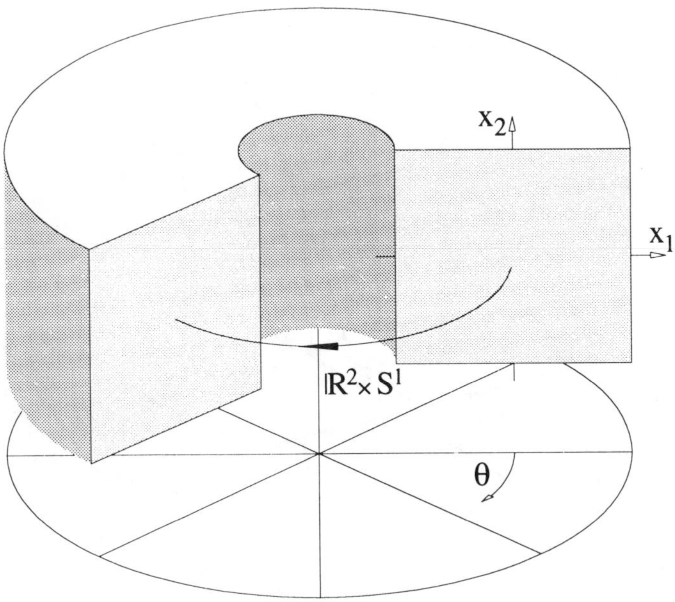

Figure 2.2: Cylindrical extended state space

For this special case of a periodic forcing the discretization time T_0 does not depend on the starting point and is naturally given by the period of excitation T . The Poincaré map in general depends also on the vector $\boldsymbol{\mu}$ of control parameters:

$$\boldsymbol{x}_{t+T} = \boldsymbol{p}\left(\boldsymbol{x}_t, \boldsymbol{\mu}\right) . \tag{2.4}$$

The map (2.4) relates to every starting point of a trajectory on the Poincaré surface (2.3) the end point which is reached by the trajectory after one excitation period T.

If one starts at a point $\boldsymbol{x}_0$ on the surface, then by iteratively applying Eq. (2.4) a stroboscopic image of the phase space can be produced. An equivalent representation is the recursion formula

$$\boldsymbol{x}_{N+1} = \boldsymbol{p}\left(\boldsymbol{x}_N, \boldsymbol{\mu}\right) \tag{2.5}$$

in terms of the discrete time $N \in \mathbb{N}$. The difference equation (2.5) describes the point map and is called time-discrete system or shortly *discrete system*. This is the basis for the investigation of the stability behavior of periodic solutions.

For periodically forced nonlinear systems closed orbits with periods which are integer multiples ℓ of the excitation period T are possible. They are called ℓ–periodic solutions of the continuous system. Figure 2.3 shows a one-periodic and a two-periodic solution in the Poincaré section.

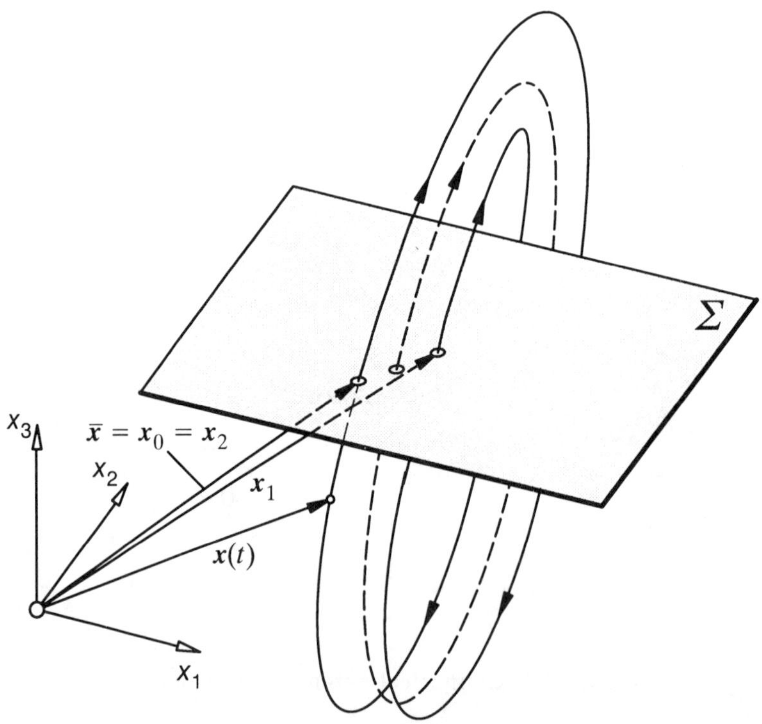

Figure 2.3: One-periodic (dashed) and two-periodic solution (solid)

The Poincaré map (2.5) converts these trajectories to a point sequence $\{\overline{\boldsymbol{x}}_1, \overline{\boldsymbol{x}}_2, \ldots, \overline{\boldsymbol{x}}_\ell\}$ which arises by an $(\ell-1)$–times application of the map with respect to a starting point $\overline{\boldsymbol{x}}_1$. The crucial point is that by a repeated application of the point map the initial point is reached again: $\boldsymbol{p}(\overline{\boldsymbol{x}}_\ell) = \overline{\boldsymbol{x}}_1$.

Such a point sequence of the discrete system is called P–ℓ solution; each point $\overline{\boldsymbol{x}}_i$ of the sequence satisfies the fixed point equation

$$\overline{\boldsymbol{x}}_i = \boldsymbol{p}^\ell\left(\overline{\boldsymbol{x}}_i, \boldsymbol{\mu}\right) \tag{2.6}$$

of the discrete system where $\boldsymbol{p}^\ell$ denotes the ℓ–times composition of the map.

The fixed points of the Poincaré map (2.4) correspond to P–1 solutions which close themselves after a single period of excitation. If $\ell \geq 2$ is valid such motions are often called *subharmonic solutions*.

2.2 Generation of a Local Symbolic Approximation

Because of the lack of an analytical mapping, analytical tools working on point maps cannot be applied. Therefore, a local approximation being valid in the neighborhood of a periodic solution respectively a fixed point $\overline{\boldsymbol{x}}$ has to be derived, Fig. 2.4.

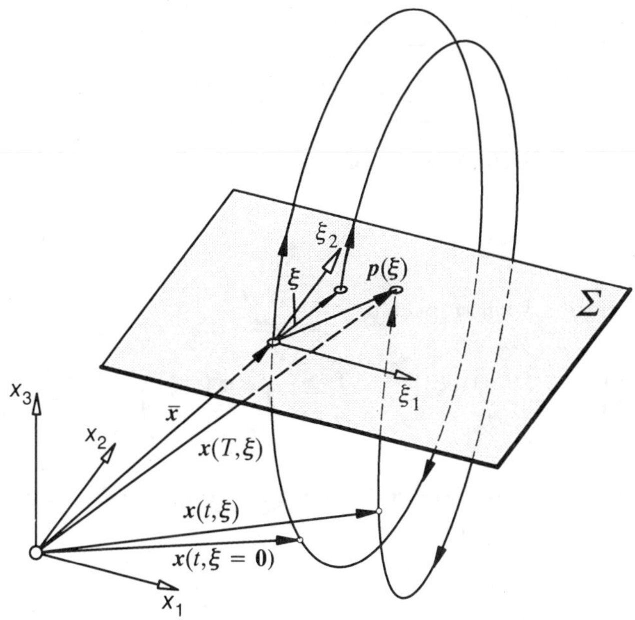

Figure 2.4: Poincaré map: periodic solution and its neighborhood

By considering $\boldsymbol{\xi}$ as deviation variable $\bar{\boldsymbol{\xi}} = \mathbf{0}$ can be chosen without restriction of the generality.

Therefore, a local approximation for $\boldsymbol{p}(\boldsymbol{\xi})$ in terms of a Taylor series expansion about the periodic solution $\bar{\boldsymbol{\xi}} = \mathbf{0}$ is formally given by

$$\boldsymbol{p}(\boldsymbol{\xi}) = \left.\frac{\partial \boldsymbol{p}(\boldsymbol{\xi})}{\partial \boldsymbol{\xi}}\right|_{\mathbf{0}} \boldsymbol{\xi} + \frac{1}{2} \left.\frac{\partial^2 \boldsymbol{p}(\boldsymbol{\xi})}{\partial \boldsymbol{\xi}^2}\right|_{\mathbf{0}} \{\boldsymbol{\xi}, \boldsymbol{\xi}\} + \frac{1}{6} \left.\frac{\partial^3 \boldsymbol{p}(\boldsymbol{\xi})}{\partial \boldsymbol{\xi}^3}\right|_{\mathbf{0}} \{\boldsymbol{\xi}, \boldsymbol{\xi}, \boldsymbol{\xi}\} + \mathcal{O}(\|\boldsymbol{\xi}\|^4) \tag{2.7}$$

where $\{\boldsymbol{\xi}, \cdots, \boldsymbol{\xi}\}$ denotes the outer product.

If $\boldsymbol{x}(t, \boldsymbol{\xi})$ is the solution of

$$\dot{\boldsymbol{x}} = \boldsymbol{f}(\boldsymbol{x}) \quad , \quad \boldsymbol{x}(t=0) = \boldsymbol{\xi} \, , \tag{2.8}$$

the Poincaré map is defined by

$$\boldsymbol{p}(\boldsymbol{\xi}) = \boldsymbol{x}(T, \boldsymbol{\xi}) \, . \tag{2.9}$$

The unknown derivative $\left.\dfrac{\partial \boldsymbol{p}(\boldsymbol{\xi})}{\partial \boldsymbol{\xi}}\right|_{\mathbf{0}}$ can be expressed as

$$\left.\frac{\partial \boldsymbol{p}(\boldsymbol{\xi})}{\partial \boldsymbol{\xi}}\right|_{\mathbf{0}} = \left.\frac{\partial \boldsymbol{x}(T, \boldsymbol{\xi})}{\partial \boldsymbol{\xi}}\right|_{\boldsymbol{\xi} = \mathbf{0}} = \left.\frac{\partial \boldsymbol{x}(t, \boldsymbol{\xi})}{\partial \boldsymbol{\xi}}\right|_{\substack{t = T \\ \boldsymbol{\xi} = \mathbf{0}}} \, . \tag{2.10}$$

The connection to the differential equation

$$\dot{\boldsymbol{x}} = \boldsymbol{f}(\boldsymbol{x}) \tag{2.11}$$

is made by differentiation with respect to t:

$$\frac{d}{dt}\left(\frac{\partial \boldsymbol{x}(t, \boldsymbol{\xi})}{\partial \boldsymbol{\xi}}\right) = \frac{\partial \boldsymbol{f}(\boldsymbol{x}(t, \boldsymbol{\xi}))}{\partial \boldsymbol{\xi}} = \frac{\partial \boldsymbol{f}(\boldsymbol{x})}{\partial \boldsymbol{x}} \left(\frac{\partial \boldsymbol{x}(t, \boldsymbol{\xi})}{\partial \boldsymbol{\xi}}\right) \, . \tag{2.12}$$

For an n–dimensional dynamical system this is a first order $(n \times n)$–matrix differential equation. With the initial conditions

$$\frac{\partial \boldsymbol{x}}{\partial \boldsymbol{\xi}}(t = 0, \boldsymbol{\xi} = \mathbf{0}) = \begin{pmatrix} 1 & & 0 \\ & \ddots & \\ 0 & & 1 \end{pmatrix} \tag{2.13}$$

numerical integration up to $t = T$ leads to

$$\frac{\partial \boldsymbol{p}}{\partial \boldsymbol{\xi}} = \frac{\partial \boldsymbol{x}}{\partial \boldsymbol{\xi}} (t = T, \boldsymbol{\xi} = \mathbf{0}) . \tag{2.14}$$

In the same manner ordinary differential equations for coefficients of higher order are supplied.

A more compact and even simpler formulation to derive the equations for the sought-after terms can be given in index notation[1]. By defining

$$_{,\widehat{i}} = \frac{\partial}{\partial \xi_i} , \quad _{,i} = \frac{\partial}{\partial x_i} , \tag{2.15}$$

first order terms are given by

$$\dot{p}_{i,\widehat{j}} = f_{i,\widehat{j}} .$$

Application of the chain rule yields

$$\dot{p}_{i,\widehat{j}} = f_{i,a} \, p_{a,\widehat{j}} . \tag{2.16}$$

For second order terms Eq. (2.16) has to be differentiated again with respect to ξ_i:

$$\dot{p}_{i,\widehat{jk}} = \left(f_{i,a} \, p_{a,\widehat{j}} \right)_{\widehat{k}} = f_{i,a\widehat{k}} \, p_{a,\widehat{j}} + f_{i,a} \, p_{a,\widehat{jk}} ,$$

$$\dot{p}_{i,\widehat{jk}} = f_{i,ab} \, p_{b,\widehat{k}} \, p_{a,\widehat{j}} + f_{i,a} \, p_{a,\widehat{jk}} . \tag{2.17}$$

The next steps can be performed equivalently.

It should be noted here that in order to avoid meaningless results for the nonlinear stability analysis at least a third order approximation of the underlying Poincaré map is required.

The linear coefficients $p_{i,j}$, the quadratic coefficients $p_{i,jk}$, and the cubic coefficients $p_{i,jkl}$ for the i–th state equation are thus given by the solution of

$$\frac{d}{dt} \begin{pmatrix} x_i \\ p_{i,j} \\ p_{i,jk} \\ p_{i,jkl} \end{pmatrix} = \begin{pmatrix} f_i \\ f_{i,a} \, p_{a,j} \\ f_{i,a} \, p_{a,jk} + f_{i,ab} \, p_{a,j} \, p_{b,k} \\ f_{i,a} \, p_{a,jkl} + f_{i,ab} \, p_{a,jk} \, p_{b,l} + f_{i,ab} \, p_{a,j} \, p_{b,kl} \\ \quad + f_{i,ab} \, p_{a,jl} \, p_{b,k} + f_{i,abc} \, p_{a,j} \, p_{b,k} \, p_{c,l} \end{pmatrix} \tag{2.18}$$

[1]Einstein's summation convention holds for the entire index formulation.

at $t = T$.[2] Here, T is the discretization time, namely the period of the solution to be studied. The only non-vanishing initial conditions are

$$p_{i,j}(t=0) = \delta_{ij} \tag{2.19}$$

where the *Kronecker–δ* is introduced. All indices a, b, c, i, j, k, l run from $1, \ldots, n$.

An approximation of the Poincaré map is not very useful if the influence of parameters cannot be studied. Therefore, parameters have to be included in a symbolic way. This is possible by a simple extension of the state space which is referred to as *suspension trick.* The elements of the parameter vector $\boldsymbol{\mu}$ under variation (control parameters) can be interpreted as state variables:

$$\begin{aligned} \dot{\boldsymbol{x}} &= \boldsymbol{f}(\boldsymbol{x}, \boldsymbol{\mu}) \,, \\ \dot{\boldsymbol{\mu}} &= \boldsymbol{0} \,. \end{aligned} \tag{2.20}$$

The parameter dependence can then be studied by means of a local Taylor series approximation of the Poincaré map of the extended system (2.20).

2.3 Implementation and Optimization

An important point is that the entire procedure can be fully automated. The whole set of ODEs (2.18) up to an approximation order of $s = 3$ for systems of arbitrary dimension n is symbolically generated by a Maple-routine as Fortran code, HAACK (1990). The solution, however, is obtained by standard numerical integration routines.

The number r of equations to be solved depends on the system dimension n and on the desired evaluation order s of the Taylor series expansion. It can be explicitly expressed by

$$r(n, s) = \sum_{i=1}^{s+1} n^i = \frac{n^{s+2} - n}{n - 1} \,. \tag{2.21}$$

Table 2.1 gives the correlation between s, n, and the number r of coupled differential equations to be derived and solved.

Table 2.1: Number r of differential equations

	n=2	n=3	n=4	n=5	n=6	n=7	n=8
s=2	14	39	84	155	258	399	584
s=3	30	120	340	780	1554	2800	4650

[2]The different quality of the indices i and $\widehat{i}$ is not indicated any more because it is only important for the correct application of the chain rule in the derivation of the equations.

From Table 2.1 it can be easily deduced that already for low-dimensional systems the task of symbolically generating the equations for the calculation of the coefficients can only be reliably performed by means of computer algebra.

By making use of the *binomial coefficient* notation[3] the number q_k of coefficients of a specified order k can be expressed by

$$q_k(n,k) = n \binom{n+k-1}{k} \tag{2.22}$$

while the number q of coefficients up to the desired evaluation order s is given by

$$q(n,s) = \sum_{k=0}^{s} q_k = n \binom{n+s}{s} . \tag{2.23}$$

This number does not coincide with the number r of equations as can be seen in Table 2.2.

Table 2.2: Number q of coefficients

	n=2	n=3	n=4	n=5	n=6	n=7	n=8
s=2	12	30	60	105	168	252	360
s=3	20	60	140	280	504	840	1320

This discrepancy is due to the fact that mixed terms occur multiple in the Taylor series expansion (2.7). Therefore, an ordering algorithm has to be applied which is also implemented in Maple.

For the computer algebra implementation it is absolutely necessary to strongly optimize the method. Otherwise, already simple problems could lead to memory overflow, compilation problems of the generated numerical code, or endless computations.

A first optimization measure is to pre-calculate all partial derivatives of the system equation (1.1). Trivial expressions for the partial derivatives are explicitly inserted into the differential equations (2.18) while non-trivial ones are separately stored. Thereby, the differential equations for the coefficients are given in terms of place-holders for the partial derivatives and thus are (almost) problem-independent.

A second measure is to make use of symmetry properties of the partial differentiation. The calculation of the partial derivatives is thus only performed with respect to a unique (ascending) sequence of state variables (given by an index sequence). The

[3] $\binom{a}{b} := \frac{a!}{b!(a-b)!}$.

corresponding partial derivative is calculated once and all during the summation procedure appearing permutations are reordered to the uniquely defined basis sequence. This leads to a considerable reduction of Maple output in terms of Fortran lines.

Example: The summation rule requires: $f_{4,121}$, $f_{4,211}$, $f_{4,112}$.

$\Rightarrow$ Only $f_{4,112} \equiv \dfrac{\partial^3 f_4}{\partial^2 x_1\, \partial x_2}$ is provided.

For an efficient calculation of partial derivatives it is also of major importance to utilize generic properties of the equations of motion of multibody systems. The symbolic vector differential equation of motion of a multibody system is given in the form

$$\boldsymbol{M}(\boldsymbol{y}, \boldsymbol{\mu}, t)\ddot{\boldsymbol{y}} + \boldsymbol{k}(\boldsymbol{y}, \dot{\boldsymbol{y}}, \boldsymbol{\mu}, t) = \boldsymbol{q}(\boldsymbol{y}, \dot{\boldsymbol{y}}, \boldsymbol{\mu}, t) \,. \tag{2.24}$$

Here, $\boldsymbol{M}$ is the $(f \times f)$ generalized mass matrix, $\boldsymbol{y}$ the $(f \times 1)$–vector of generalized coordinates, $\boldsymbol{k}$ the $(f \times 1)$–vector of gyroscopic and centrifugal forces, $\boldsymbol{q}$ the $(f \times 1)$–vector of generalized applied forces, and f the number of degrees of freedom, Kreuzer and Schiehlen (1990), Kreuzer and Leister (1991).

The first order state equation has the following general structure:

$$\underbrace{\frac{d}{dt}\begin{bmatrix} \boldsymbol{y} \\ \dot{\boldsymbol{y}} \\ \boldsymbol{\mu} \end{bmatrix}}_{\dot{\boldsymbol{x}}} = \underbrace{\begin{bmatrix} \dot{\boldsymbol{y}} \\ \boldsymbol{M}^{-1}(\boldsymbol{q} - \boldsymbol{k}) \\ \boldsymbol{0} \end{bmatrix}}_{\boldsymbol{f}(\boldsymbol{x})} \tag{2.25}$$

where $\boldsymbol{M}^{-1}$ can be expressed by

$$\boldsymbol{M}^{-1} = \frac{1}{\det(\boldsymbol{M})} \operatorname{adj}(\boldsymbol{M}) \,. \tag{2.26}$$

The only non-trivial equation for the acceleration is given by

$$\ddot{\boldsymbol{y}} = \frac{1}{s(\boldsymbol{y}, \boldsymbol{\mu}, t)}\, \boldsymbol{A}(\boldsymbol{y}, \boldsymbol{\mu}, t)\, \boldsymbol{v}(\boldsymbol{y}, \dot{\boldsymbol{y}}, \boldsymbol{\mu}, t) \equiv \boldsymbol{f}_a(\boldsymbol{y}, \dot{\boldsymbol{y}}, \boldsymbol{\mu}, t) \tag{2.27}$$

with a scalar $s = \det \boldsymbol{M}$, an $(f \times f)$–matrix $\boldsymbol{A} = \operatorname{adj} \boldsymbol{M}$, and an $(f \times 1)$–vector $\boldsymbol{v} = \boldsymbol{q} - \boldsymbol{k}$. The calculation of the partial derivatives of the equation $\boldsymbol{f}_a(\boldsymbol{y}, \dot{\boldsymbol{y}}, \boldsymbol{\mu}, t)$ with respect to the state variables $\boldsymbol{y}$ and $\dot{\boldsymbol{y}}$ and the control parameters $\boldsymbol{\mu}$ is performed via generally applicable computer algebra tools in two steps:

1. All non-trivial partial derivatives of the explicitly evaluated problem-dependent quantities s, $\boldsymbol{v}$, and $\boldsymbol{A}$ are pre-calculated and separately stored.

2. All partial derivatives of Eq. (2.27) with its implicitly given (and hence problem-independent) dependencies are formally calculated in terms of the quantities calculated in Step 1 by successively applying the chain rule up to the desired order. Trivial expressions calculated in Step 1 are explicitly inserted.

For the numerical evaluation of the partial derivatives the stored symbolic relations of the partial derivatives calculated in Step 1 are once evaluated and then used repetitively for determining the entire expressions.

3 Stability of Periodic Orbits and System Reduction

The stability of periodic solutions can be determined for the non-critical case by the linear part of the Poincaré map which is described by the so-called *monodromy matrix*. For the critical case the linearly decoupled system can be reduced to a lower dimensional critical subsystem via center manifold theory. Then, the nonlinear terms have to be considered for the stability analysis.

3.1 Stability Boundaries

The periodic orbit is stable if the eigenvalues of the respective monodromy matrix (which are also called *characteristic* or *Floquet multipliers*) lay inside the unit circle ($|\lambda_i| < 1$), otherwise it is unstable ($|\lambda_i| > 1$) or critical ($|\lambda_i| = 1$).

As long as a system does not have critical eigenvalues it is *structurally stable*. Through variations of the parameters the solution manifolds change continuously and differentiably. But if one or more critical eigenvalues occur, then for quasistationary parameter variations the topological structure of the phase curves changes qualitatively; the system undergoes a so-called *bifurcation*.

If the system approaches the stability boundary due to a single parameter variation, for bifurcations of discrete systems three different generic cases can occur (see Fig. 3.1):

1. Saddle-node, transcritical, or pitchfork bif.: $\lambda = +1$
2. Flip bifurcation : $\lambda = -1$
3. Hopf bifurcation : $\lambda = a \pm j\,b\,,\; a^2 + b^2 = 1\,,\; b \neq 0$

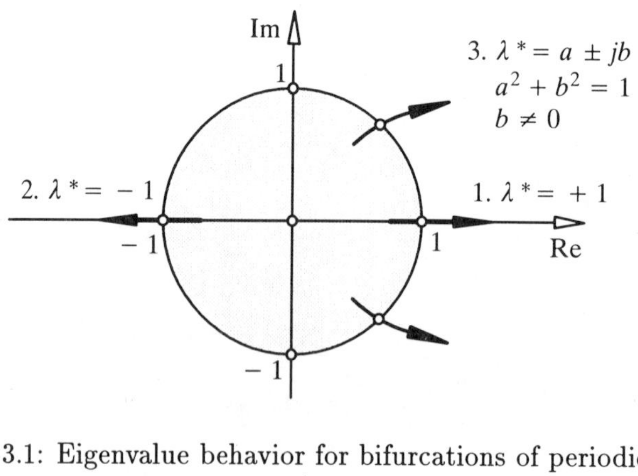

Figure 3.1: Eigenvalue behavior for bifurcations of periodic orbits

3.2 Detection of Bifurcation Points

On the basis of the linearized map with symbolically included parameter dependence it is possible to give an estimation of the critical parameter value for which a bifurcation takes place. The starting point is the symbolically given map $\boldsymbol{p}(\boldsymbol{\xi}, \boldsymbol{\mu})$. The following steps lead to the complete problem formulation as is given in Table 3.1:

- Calculation of the *Jacobian*: $\boldsymbol{J}(\boldsymbol{\xi}, \boldsymbol{\mu}) = \partial \boldsymbol{p} / \partial \boldsymbol{\xi}$.
- Calculation of the *characteristic equation* (Section 3.2): $P(\boldsymbol{\xi}, \boldsymbol{\mu}, \lambda) = 0$.
- Substituting $\lambda = a + j\, b$.

Table 3.1: Complete problem formulation to detect bifurcation points

1	Real part condition	:	$\mathrm{Re}\,[P(\boldsymbol{\xi}, \boldsymbol{\mu}, a, b)]$	=	0
1	Imaginary part condition	:	$\mathrm{Im}\,[P(\boldsymbol{\xi}, \boldsymbol{\mu}, a, b)]$	=	0
1	Criticality condition	:	$a^2 + b^2 - 1$	=	0
n	Fixed point conditions	:	$\boldsymbol{\xi} - \boldsymbol{p}(\boldsymbol{\xi}, \boldsymbol{\mu})$	=	$\boldsymbol{0}$
$n_r - 1$	Parameter conditions	:	$\boldsymbol{\delta}(\boldsymbol{\mu})$	=	$\boldsymbol{0}$

Computer algebra is used to set up the equations which are solved numerically according to the following iterative procedure:

1. Select the parameter in the vicinity of a bifurcation point (e.g. given by a global analysis via Ljapunov exponents, Kreuzer (1987)).
2. Calculate the fixed point of the Poincaré map.
3. Derive a parametrized approximation of the Poincaré map.
4. Calculate a correction of the bifurcation point in parameter and state space using the equations given in Table 3.1.
5. Continue iteratively until the bifurcation point is reached within a certain tolerance for which the system is considered to be at the stability boundary with critical values as given below:

$$\left.\begin{matrix} a^\star \\ b^\star \end{matrix}\right\} \quad \text{Critical eigenvalue } \lambda^\star = a^\star + j\, b^\star ,$$

$\boldsymbol{\xi}^\star$: Critical fixed point $(n \times 1)$,

$\boldsymbol{\mu}^\star$: Critical parameter vector $(n_r \times 1)$.

3.3 Linear Decoupling

For the application of the center manifold reduction the system equations have to be decoupled with respect to their critical and uncritical part. The decoupling leads to the *eigenvalue problem*

$$\boldsymbol{J}\,\boldsymbol{x} = \lambda\,\boldsymbol{x} \tag{3.1}$$

where $\boldsymbol{J}$ is an arbitrary but real $(n \times n)$–matrix. The $(n \times 1)$–vector $\boldsymbol{x}$ and the scalar parameter λ are still undetermined. Equation (3.1) can be written as well as a homogenious set of equations:

$$\boldsymbol{F}(\lambda)\,\boldsymbol{x} = (\,\boldsymbol{J} - \lambda\,\boldsymbol{I}\,)\,\boldsymbol{x} = \boldsymbol{0} \quad \text{with} \quad \boldsymbol{x} \neq \boldsymbol{0}\,. \tag{3.2}$$

Here, $\boldsymbol{I}$ is the $(n \times n)$–unit matrix. The $(n \times n)$–Matrix $\boldsymbol{F}(\lambda)$ is the so-called *characteristic matrix* corresponding to $\boldsymbol{J}$. Equation (3.2) has non-trivial solutions $\boldsymbol{x}_i$ if and only if the determinant of the characteristic matrix vanishes, i.e.

$$P(\lambda) \equiv \det\left[\,\boldsymbol{F}(\lambda)\,\right] = 0\,. \tag{3.3}$$

This characteristic equation is also called *characteristic polynomial* of the matrix $\boldsymbol{J}$ and has — as a real equation of order n in λ — exactly n (real or conjugate-complex) roots λ_i, $i = 1, \ldots, n$. Only for this roots λ_i — the so-called *eigenvalues* or *characteristic numbers* of the matrix $\boldsymbol{J}$ — the eigenvalue problem (3.1) has non-trivial solutions $\boldsymbol{x}_i$, $i = 1, \ldots, n$, the so-called *eigensolutions* or *eigenvectors* (modal columns) of $\boldsymbol{J}$.

The most important point is to provide the eigenvectors $\boldsymbol{x}_i$ respectively the modal matrix $\boldsymbol{X}$ whose columns are in the simplest case given by the eigenvectors.

By means of a linear coordinate transformation

$$\boldsymbol{x} = \boldsymbol{X}\boldsymbol{w}\,, \tag{3.4}$$

the original system

$$\boldsymbol{x}_{N+1} = \boldsymbol{J}\,\boldsymbol{x}_N + \boldsymbol{P}(\boldsymbol{x}_N) \tag{3.5}$$

can be transformed into a linearly decoupled system:

$$\boldsymbol{w}_{N+1} = \underbrace{\boldsymbol{X}^{-1}\boldsymbol{J}\boldsymbol{X}}_{\boldsymbol{\Lambda}}\,\boldsymbol{w}_N + \boldsymbol{X}^{-1}\boldsymbol{P}(\boldsymbol{X}\boldsymbol{w}_N)\,. \tag{3.6}$$

Thereby, $\boldsymbol{\Lambda}$ is the diagonal matrix of the eigenvalues of $\boldsymbol{J}$.

The characteristic polynomial of an $(n \times n)$–matrix $\boldsymbol{J}$ does not necessarily lead to n distinct eigenvalues λ_i because multiple roots are also possible. The s distinct eigenvalues λ_j, $j = 1, \ldots, s$ are distinguished by their *algebraic multiplicity* v_j . The number of eigenvectors corresponding to an eigenvalue λ_j is in general not equal to v_j but to the *geometric multiplicity* or *defect* d_j which is given by

$$d_j = n - \operatorname{rank}\left[\boldsymbol{F}(\lambda_j)\right], \quad 1 \leq d_j \leq v_j \,. \tag{3.7}$$

The case $d_j < v_j$ leads to a lack of $v_j - d_j$ eigenvectors, i.e. columns in the modal matrix. In order to provide additional vectors for this missing columns one needs the so-called *principal vectors* corresponding to the eigenvalue. However, this topic will not be discussed here.

Furthermore, it can be shown that even for the case of conjugate-complex eigenvalues a block decoupling can be achieved by exclusively applying real transformations.

Modern computer algebra systems provide extensive facilities for vector and matrix manipulation. The symbolic computation of eigenvalues and eigenvectors, however, and thus the treatment of the linear decoupling problem by computer algebra is not always possible for matrices with pure symbolic elements. Already for very low system dimensions ($n \geq 3$) the results for the eigenvalues become so extensive and unhandy that it is scarcely efficient or even impossible to proceed. Experience shows that it is especially for higher-dimensional systems much more efficient to handle matrices with exclusively numerical elements externally by numerical routines than by numerical Maple-procedures.

The computer-aided solution of the decoupling problem can be summarized as follows:

- The system dimension is generally the limiting factor for handling the eigenvalue problem by means of symbolic computation.
- For purely numerical matrix elements the eigenvalue problem should always be treated by numerical methods.
- If a nonlinear system (3.5) is to be linearly decoupled the transformation process itself — with known modal matrix $\boldsymbol{X}$ — must in any case be symbolically performed.

3.4 Reduction to a Critical Subsystem

Center manifold theory provides a very useful method for the simplification of critical systems. After the so-called *center manifold reduction* the stability analysis has to be performed only for a lower-dimensional critical subsystem. The dimension of this *bifurcation system* is equal to the number of critical eigenvalues.

Suppose the linearly decoupled system is given by

$$\begin{aligned} \boldsymbol{u}_{N+1} &= \boldsymbol{J}^c \boldsymbol{u}_N + \boldsymbol{P}^c(\boldsymbol{u}_N, \boldsymbol{v}_N), \quad & |\lambda_i[\boldsymbol{J}^c]| = 1 \,\forall\, i, \\ \boldsymbol{v}_{N+1} &= \boldsymbol{J}^s \boldsymbol{v}_N + \boldsymbol{P}^s(\boldsymbol{u}_N, \boldsymbol{v}_N), \quad & |\lambda_j[\boldsymbol{J}^s]| < 1 \,\forall\, j, \end{aligned} \tag{3.8}$$

where

$$\begin{aligned} \boldsymbol{P}^c(\boldsymbol{0},\boldsymbol{0}) &= \boldsymbol{0}, \qquad & D\boldsymbol{P}^c(\boldsymbol{0},\boldsymbol{0}) &= \boldsymbol{0}, \\ \boldsymbol{P}^s(\boldsymbol{0},\boldsymbol{0}) &= \boldsymbol{0}, \qquad & D\boldsymbol{P}^s(\boldsymbol{0},\boldsymbol{0}) &= \boldsymbol{0}, \end{aligned}$$

with critical variables $\boldsymbol{u} \in \mathbb{R}^{n_c}$ and uncritical variables $\boldsymbol{v} \in \mathbb{R}^{n_s}$. The $(n_c \times n_c)$–matrix $\boldsymbol{J}^c$ and the $(n_s \times n_s)$–matrix $\boldsymbol{J}^s$ determine the linearly decoupled part of the map. Their eigenvalue spectrum leads to the distinction between a critical and stable part[4]. Similarly, the functions $\boldsymbol{P}^c, \boldsymbol{P}^s \in C^p$, $(p \geq 2)$ denote the nonlinear parts of the respective subsystems.

An invariant manifold is called a *center manifold* W^c for (3.8) if it can be locally represented as

$$W^c = \{\, (\boldsymbol{u},\boldsymbol{v}) \in \mathbb{R}^{n_c} \times \mathbb{R}^{n_s} \mid \boldsymbol{v} = \boldsymbol{h}(\boldsymbol{u}),\ \|\boldsymbol{u}\| < \delta,\ \boldsymbol{h}(\boldsymbol{0}) = \boldsymbol{0},\ D\boldsymbol{h}(\boldsymbol{0}) = \boldsymbol{0} \,\} \tag{3.9}$$

for δ being sufficiently small. Note that the stable variables $\boldsymbol{v}$ are substituted by the function $\boldsymbol{h}$ of the critical variables $\boldsymbol{u}$ which is used to 'decouple' the critical equations from the unstable ones.

The center manifold W^c is in the fixed point $\overline{\boldsymbol{x}}$ tangential to the linear eigenspace E^c spanned by the critical eigenvectors, Fig. 3.2.

The following two theorems proven in the book by CARR (1981) state the existence of a center manifold and the relation of the stability behavior of the reduced system to that of the original system:

Theorem 3.1 *There exists a C^p center manifold for (3.8). The dynamics of (3.8) restricted to the center manifold is given (for sufficiently small $\boldsymbol{z}$) by the n_c–dimensional map*

$$\boldsymbol{z}_{N+1} = \boldsymbol{J}^c \boldsymbol{z}_N + \boldsymbol{P}^c(\boldsymbol{z}_N, \boldsymbol{h}(\boldsymbol{z}_N)) . \tag{3.10}$$

Theorem 3.2 *Suppose the zero solution of (3.10) is stable (unstable). Then the zero solution of (3.8) is stable (unstable).*

[4]The assumption of an additional unstable part does not lead to any serious restrictions, WIGGINS (1990).

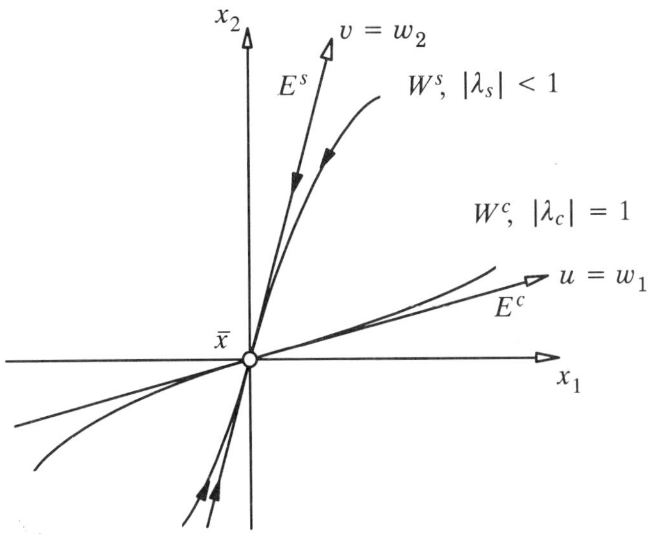

Figure 3.2: Invariant manifolds and canonical coordinates

The condition for the center manifold can be derived as follows:

$$\begin{aligned}
\boldsymbol{h}(\boldsymbol{u}_N) &= \boldsymbol{v}_N\,, \\
\boldsymbol{h}(\boldsymbol{u}_{N+1}) &= \boldsymbol{v}_{N+1}\,, \\
\boldsymbol{h}(\boldsymbol{J}^c\,\boldsymbol{u}_N + \boldsymbol{P}^c(\boldsymbol{u}_N, \boldsymbol{v}_N)) &= \boldsymbol{J}^s\,\boldsymbol{v}_N + \boldsymbol{P}^s(\boldsymbol{u}_N, \boldsymbol{v}_N)\,, \\
\boldsymbol{h}(\boldsymbol{J}^c\,\boldsymbol{u}_N + \boldsymbol{P}^c(\boldsymbol{u}_N, \boldsymbol{h}(\boldsymbol{u}_N))) &= \boldsymbol{J}^s\,\boldsymbol{h}(\boldsymbol{u}_N) + \boldsymbol{P}^s(\boldsymbol{u}_N, \boldsymbol{h}(\boldsymbol{u}_N))\,.
\end{aligned}$$

and so

$$\Phi(\boldsymbol{h}(\boldsymbol{u}_N)) := \boldsymbol{h}(\boldsymbol{J}^c\,\boldsymbol{u}_N + \boldsymbol{P}^c(\boldsymbol{u}_N, \boldsymbol{h}(\boldsymbol{u}_N))) - \boldsymbol{J}^s\,\boldsymbol{h}(\boldsymbol{u}_N) - \boldsymbol{P}^s(\boldsymbol{u}_N, \boldsymbol{h}(\boldsymbol{u}_N)) = \boldsymbol{0}\,. \quad (3.11)$$

The last theorem by CARR (1981) is crucial for computing an approximation of the center manifold and enables one to use computer algebra for doing this, RAND AND ARMBRUSTER (1987):

Theorem 3.3 *Let* $\boldsymbol{\varphi} : \mathbb{R}^{n_c} \to \mathbb{R}^{n_s}$ *be a* $\mathcal{C}^1$ *map with* $\boldsymbol{\varphi}(\boldsymbol{0}) = \boldsymbol{0}$, $D\boldsymbol{\varphi}(\boldsymbol{0}) = \boldsymbol{0}$ *and* $\Phi(\boldsymbol{\varphi}(\boldsymbol{u})) = \mathcal{O}(\|\boldsymbol{u}\|^q)$ *as* $\boldsymbol{u} \to \boldsymbol{0}$ *for some* $q > 1$. *Then*

$$\boldsymbol{h}(\boldsymbol{u}) = \boldsymbol{\varphi}(\boldsymbol{u}) + \mathcal{O}(\|\boldsymbol{u}\|^q) \qquad \text{as} \quad \boldsymbol{u} \to \boldsymbol{0}\,.$$

If $\boldsymbol{h}(\boldsymbol{u})$ is substituted into (3.11) by a multivariate power series $\boldsymbol{\varphi}(\boldsymbol{u})$ with unknown coefficients an order calculation up to the desired evaluation order leads to an approximation $\boldsymbol{\varphi}(\boldsymbol{u})$ of $\boldsymbol{h}(\boldsymbol{u})$.

3.5 Center Manifolds and Near Identity Transformations

In the theory of normal forms (Chapter 4) certain coordinate changes (so-called *near identity transformations*) are used to simplify the system equations to their normal form. The same type of coordinate transformations can be used for the calculation of center manifolds, FREIRE ET AL. (1989):

The near identity transformation

$$\begin{pmatrix} \boldsymbol{u}_N \\ \boldsymbol{v}_N \end{pmatrix} \rightarrow \begin{pmatrix} \boldsymbol{z}_N \\ \boldsymbol{y}_N \end{pmatrix} + \begin{pmatrix} \boldsymbol{0} \\ \boldsymbol{h}(\boldsymbol{z}_N) \end{pmatrix} \tag{3.12}$$

with $\boldsymbol{z} \in \mathrm{IR}^{n_c}$, $\boldsymbol{y} \in \mathrm{IR}^{n_s}$, $\boldsymbol{h}(\boldsymbol{0}) = \boldsymbol{0}$, $D\boldsymbol{h}(\boldsymbol{0}) = \boldsymbol{0}$ changes (3.8) to

$$\begin{aligned} \boldsymbol{z}_{N+1} &= \boldsymbol{J}^c \boldsymbol{z}_N + \boldsymbol{Q}^c(\boldsymbol{z}_N, \boldsymbol{y}_N) \,, \\ \boldsymbol{y}_{N+1} &= \boldsymbol{J}^s \boldsymbol{y}_N + \boldsymbol{Q}^s(\boldsymbol{z}_N, \boldsymbol{y}_N) \,, \end{aligned} \tag{3.13}$$

where (with dropped index N)

$$\begin{aligned} \boldsymbol{Q}^c(\boldsymbol{z}_N, \boldsymbol{y}_N) &= \boldsymbol{P}^c(\boldsymbol{z}_N, \boldsymbol{y}_N + \boldsymbol{h}(\boldsymbol{z}_N)) \,, \\ \boldsymbol{Q}^s(\boldsymbol{z}_N, \boldsymbol{y}_N) &= \boldsymbol{J}^s \boldsymbol{h}(\boldsymbol{z}_N) + \boldsymbol{P}^s(\boldsymbol{z}_N, \boldsymbol{y}_N + \boldsymbol{h}(\boldsymbol{z}_N)) \\ &\quad - \boldsymbol{h}(\boldsymbol{J}^c \boldsymbol{z}_N + \boldsymbol{P}^c(\boldsymbol{z}_N, \boldsymbol{y}_N + \boldsymbol{h}(\boldsymbol{z}_N))) \,. \end{aligned} \tag{3.14}$$

By choosing $\boldsymbol{h}(\boldsymbol{z})$ such that $\boldsymbol{y} \equiv \boldsymbol{0}$ (which is the same as $\boldsymbol{Q}^s(\boldsymbol{z}, \boldsymbol{0}) \equiv \boldsymbol{0}$) one can deduce that $\boldsymbol{h}(\boldsymbol{z})$ must verify Eq. (3.11).

The reduced system becomes

$$\boldsymbol{z}_{N+1} = \boldsymbol{J}^c \boldsymbol{z}_N + \boldsymbol{Q}^c(\boldsymbol{z}_N, \boldsymbol{0}) = \boldsymbol{J}^c \boldsymbol{z}_N + \boldsymbol{P}^c(\boldsymbol{z}_N, \boldsymbol{h}(\boldsymbol{z}_N)) \,. \tag{3.15}$$

The conclusion is: center manifold calculation is equivalent to performing a near identity transformation on the system with the condition $\boldsymbol{y} \equiv \boldsymbol{0}$.

3.6 Parameter Dependence

The parameter dependence must be taken into account for the center manifold approach by again regarding the parameters as trivial state variables, see Chapter 2. Suppose the system is written as

$$\begin{aligned} \boldsymbol{u}_{N+1} &= \boldsymbol{J}^c \boldsymbol{u}_N + \boldsymbol{P}^c(\boldsymbol{u}_N, \boldsymbol{v}_N, \boldsymbol{\mu}_N) \,, \quad |\, \lambda_i[\boldsymbol{J}^c] \,| = 1 \;\forall\, i \,, \\ \boldsymbol{\mu}_{N+1} &= \boldsymbol{\mu}_N \,, \\ \boldsymbol{v}_{N+1} &= \boldsymbol{J}^s \boldsymbol{v}_N + \boldsymbol{P}^s(\boldsymbol{u}_N, \boldsymbol{v}_N, \boldsymbol{\mu}_N) \,, \quad |\, \lambda_j[\boldsymbol{J}^s] \,| < 1 \;\forall\, j \,, \end{aligned} \tag{3.16}$$

where

$$\boldsymbol{P}^c(\mathbf{0},\mathbf{0},\mathbf{0}) = \mathbf{0}\,, \qquad D\boldsymbol{P}^c(\mathbf{0},\mathbf{0},\mathbf{0}) = \mathbf{0}\,,$$
$$\boldsymbol{P}^s(\mathbf{0},\mathbf{0},\mathbf{0}) = \mathbf{0}\,, \qquad D\boldsymbol{P}^s(\mathbf{0},\mathbf{0},\mathbf{0}) = \mathbf{0}\,,$$

with the parameter vector $\boldsymbol{\mu} \in \mathrm{IR}^{n_r}$ [5]. What is gained from this proceeding is that the center manifold theory can be applied just in the same way as was outlined before (with additional critical variables $\boldsymbol{\mu}$). The existence of the center manifold in a small neighborhood of $(\boldsymbol{u},\boldsymbol{\mu}) = (\mathbf{0},\mathbf{0})$ is guaranteed and therefore all bifurcating solutions (which can arise by variations of the parameters) are contained in the lower-dimensional reduced system, WIGGINS (1990). The flow on the center manifold is then given by

$$\begin{aligned} \boldsymbol{z}_{N+1} &= \boldsymbol{J}^c \boldsymbol{z}_N + \boldsymbol{P}^c(\boldsymbol{z}_N, \boldsymbol{h}(\boldsymbol{z}_N,\boldsymbol{\mu}_N), \boldsymbol{\mu}_N)\,, \\ \boldsymbol{\mu}_{N+1} &= \boldsymbol{\mu}_N\,. \end{aligned} \tag{3.17}$$

Analogously to Eq. (3.11) the explicit calculation of the center manifold now becomes

$$\begin{aligned} &\boldsymbol{h}(\boldsymbol{J}^c \boldsymbol{u}_N + \boldsymbol{P}^c(\boldsymbol{u}_N, \boldsymbol{h}(\boldsymbol{u}_N,\boldsymbol{\mu}), \boldsymbol{\mu})) \\ &\quad - \boldsymbol{J}^s \boldsymbol{h}(\boldsymbol{u}_N,\boldsymbol{\mu}) - \boldsymbol{P}^s(\boldsymbol{u}_N, \boldsymbol{h}(\boldsymbol{u}_N,\boldsymbol{\mu}), \boldsymbol{\mu}) = \mathbf{0}\,. \end{aligned} \tag{3.18}$$

[5]The matrices $\boldsymbol{J}^c$ and $\boldsymbol{J}^s$ do not depend on the parameters any more because all terms which include parameters now become nonlinear and are contained in $\boldsymbol{P}^c$ and $\boldsymbol{P}^s$.

4 Normal Forms

Corresponding to every critical subsystem one can find simpler systems — called normal forms — which are identical in the linear part but contain only so-called 'resonant' nonlinear terms. For the determination of the resonant terms near identity transformations are performed. Bifurcation problems can be simplified to such a local normal form representation. This allows — with a proper choice of the normal form basis — a classification and stability analysis of the considered bifurcation.

4.1 Transformation Procedure

The starting point for the following considerations is the flow on the center manifold [6] of an ordinary system of difference equations (3.8):

$$\boldsymbol{z}_{N+1} = \boldsymbol{p}(\boldsymbol{z}_N) = \boldsymbol{J}\,\boldsymbol{z}_N + \boldsymbol{P}(\boldsymbol{z}_N) \tag{4.1}$$

The state vector $\boldsymbol{z}$ is contained in the center subspace and all eigenvalues of the Jacobian $\boldsymbol{J}$ are on the unit circle. In order to examine the system behavior in the neighborhood of the fixed point the multivariate Taylor expansion of (4.1) about the origin has to be considered:

$$\boldsymbol{z}_{N+1} = \boldsymbol{J}\,\boldsymbol{z}_N + \boldsymbol{P}_2(\boldsymbol{z}_N) + \ldots + \boldsymbol{P}_s(\boldsymbol{z}_N) + \mathcal{O}\left(\|\boldsymbol{z}\|^{s+1}\right). \tag{4.2}$$

The nonlinear terms of the multivariate Taylor series expansion are *vectorial polynomials*, i.e. vector fields with polynomial components.

Vector Space Formalism for Multivariate Polynomials

Polynomials can be decomposed into *monomials* which again can be represented uniquely as n–tupels $\boldsymbol{m}$ of the exponents of the variables. The following notation is used:

$$\boldsymbol{z}^{\boldsymbol{m}} = \prod_{i=1}^{n} z_i^{m_i} = z_1^{m_1}\, z_2^{m_2} \cdots z_n^{m_n}\,. \tag{4.3}$$

The sum over the components m_i of an exponent tupel is called *order* k:

$$k = \sum_{i=1}^{n} m_i\,. \tag{4.4}$$

[6]In the following, the index c for the critical subsystem is omitted for brevity; the dimension of the critical subsystem will therefore be denoted by n.

The number of possibilities to distribute the numbers $0, 1, \ldots, k$ with sum k over n positions under consideration of succesion is given by

$$\hat{l}_k = \hat{l}\,(k, n) = \binom{n+k-1}{k} . \tag{4.5}$$

That means that to every finite order k and dimension n there exist exactly $\hat{l}_k$ different monomials $z^{\boldsymbol{m}(k,j)}$, $j = 1, \ldots, \hat{l}_k$.

Table 4.1 supplies the exponent tupels for the lowest-order terms of evaluation for two-dimensional systems.

Table 4.1: Exponent tupels for system dimension two

$n=2$	$k=2$	$k=3$	$k=4$	$k=5$
$\boldsymbol{m}(k,1)$	[2, 0]	[3, 0]	[4, 0]	[5, 0]
$\boldsymbol{m}(k,2)$	[1, 1]	[2, 1]	[3, 1]	[4, 1]
$\boldsymbol{m}(k,3)$	[0, 2]	[1, 2]	[2, 2]	[3, 2]
$\boldsymbol{m}(k,4)$	–	[0, 3]	[1, 3]	[2, 3]
$\boldsymbol{m}(k,5)$	–	–	[0, 4]	[1, 4]
$\boldsymbol{m}(k,6)$	–	–	–	[0, 5]

With this notation the terms according to an evaluation order k of (4.2) can be written shortly as:

$$\boldsymbol{P}_k(\boldsymbol{z}) = \sum_{i=1}^{n} \sum_{j=1}^{\hat{l}_k} a_{k,i,j}\; \boldsymbol{z}^{\boldsymbol{m}(k,j)}\; \boldsymbol{e}_i . \tag{4.6}$$

In the center subspace the unit vectors $\boldsymbol{e}_i$ form an orthogonal coordinate system $K\{\boldsymbol{0}, \boldsymbol{e}_1, \ldots, \boldsymbol{e}_n\}$. The evaluation coefficients $a_{k,i,j}$ are given by the partial derivatives of the right hand side $\boldsymbol{p}$ with respect to the variables $\boldsymbol{z}$ about the evaluation point $\overline{\boldsymbol{z}} = \boldsymbol{0}$:

$$a_{k,i,j} = \frac{1}{m_1(k,j)! \cdots m_n(k,j)!} \left. \frac{\partial^k p_i(\boldsymbol{z})}{\partial z_1^{m_1(k,j)} \cdots \partial z_n^{m_n(k,j)}} \right|_{\boldsymbol{z}=\boldsymbol{0}} . \tag{4.7}$$

The n–dimensional vector polynomials of order k form the linear vector space $\mathcal{H}_{k,n}$. Firstly, this space is considered over the field of real numbers ($a_{k,i,j} \in \mathbb{R}$). Therefore, the real coefficients $a_{k,i,j}$ of the multivariate Taylor expansion can also be regarded as coordinates with respect to a basis of the vector space $\mathcal{H}_{k,n}$.

The natural basis is given by the $n \cdot \hat{l}_k$ *vector monomials* which result in all possible combinations of the positive and integer exponent tupels $\boldsymbol{m}(k,j)$ with the unit vectors $\boldsymbol{e}_i$. If the vector monomials of this basis are ordered with ascending index i of the unit vectors it follows the general result:

$$\mathcal{H}_{k,n} = \left\{ \boldsymbol{z}^{\boldsymbol{m}(k,1)}\, \boldsymbol{e}_1 \,, \,\dots\,, \boldsymbol{z}^{\boldsymbol{m}(k,\hat{l}_k)}\, \boldsymbol{e}_1 \,, \boldsymbol{z}^{\boldsymbol{m}(k,1)}\, \boldsymbol{e}_2 \,, \,\dots\,, \boldsymbol{z}^{\boldsymbol{m}(k,\hat{l}_k)}\, \boldsymbol{e}_n \right\} . \tag{4.8}$$

The use of a threefold indexing of the coefficients of the multivariate Taylor expansion can be avoided by changing the indices i,j into a single index l [7]: this index l of the evaluation coefficients runs through all $n \cdot \hat{l}_k$ basis vectors of $\mathcal{H}_{k,n}$ and is uniquely given by the indices i,j :

$$l = (i-1) * \hat{l}_k + j \,, \quad i = 1 \dots n \quad \text{und} \quad j = 1 \dots \hat{l}_k \,. \tag{4.9}$$

To change the indexing back the following equations can be used:

$$\begin{aligned} j &= 1 + (l-1) \mod \hat{l}_k \,, \\ i &= 1 + \frac{l-j}{\hat{l}_k} \,. \end{aligned} \tag{4.10}$$

Using (4.9) the coefficients $a_{k,i,j}$ of the evaluation order k can be rewritten as so-called *coefficient vectors* $\boldsymbol{\alpha}_k$ in the space of vector polynomials of dimension n and order k :

$$\boldsymbol{\alpha}_k = \left(\alpha_{k,1} \; \dots \; \alpha_{k,n\cdot\hat{l}_k} \right)^T \equiv \left(a_{k,1,1} \; a_{k,1,2} \; \dots \; a_{k,1,\hat{l}_k} \; a_{k,2,1} \; \dots \; a_{k,n,\hat{l}_k} \right)^T . \tag{4.11}$$

In the following computations and examples the indexing is chosen with regard to the simplest description. It is always possible to transform the results from one case to the other by using Eqs. (4.9) and (4.10).

As an example for a two-dimensional vector field with the state vector $\boldsymbol{z} = (z_1 \; z_2)^T$ the Taylor series expansion about the origin up to second order will be performed for the system

$$\begin{aligned} p_1(\boldsymbol{z}) &= z_2 + \sin(z_1 z_2) \\ p_2(\boldsymbol{z}) &= -z_1 \,. \end{aligned}$$

The first partial derivatives of this vector field with respect to the state variables evaluated for the origin are given by the Jacobian:

[7] The first index of the coefficients is always the evaluation order k . In the case when the single l–indexing is used, the coefficients are denoted by small greek letters $\alpha, \beta, \dots$; in the other case, when the (i,j)–indexing is used, by small latin letters $a, b, \dots$

$$J = \begin{pmatrix} 0 & 1 \\ -1 & 0 \end{pmatrix} .$$

By inserting the coordinates of the origin $(z_1 \; z_2)^T = (0 \; 0)^T$ into the second order partial derivatives one obtains the coefficients of the multivariate Taylor series expansion:

$$\begin{aligned} a_{2,1,1} &= 0 \; , \; a_{2,1,2} = 1 \; , \; a_{2,1,3} = 0 \; , \\ a_{2,2,1} &= 0 \; , \; a_{2,2,2} = 0 \; , \; a_{2,2,3} = 0 \; . \end{aligned}$$

The second-order coefficients can also be represented as a vector $\boldsymbol{\alpha}_2$:

$$\boldsymbol{\alpha}_2 = (0\,1\,0\,0\,0\,0)^T .$$

The vector monomial with the first unit vector $(1\;0)^T$ and the second exponent tupel $\boldsymbol{m}(2) = [1,1]$ corresponds to the only relevant coefficient $a_{2,1,2}$:

$$\boldsymbol{z}^{\boldsymbol{m}(2,2)}\boldsymbol{e}_1 = \begin{pmatrix} z_1 z_2 \\ 0 \end{pmatrix} .$$

Consequently, the Taylor series evaluation up to second order follows:

$$\boldsymbol{p}(\boldsymbol{z}) = \begin{pmatrix} 0 & 1 \\ -1 & 0 \end{pmatrix} \begin{pmatrix} z_1 \\ z_2 \end{pmatrix} + \begin{pmatrix} z_1 z_2 \\ 0 \end{pmatrix} + \mathcal{O}\left(\|\boldsymbol{z}\|^3\right) .$$

In the general two-dimensional case the natural basis of the vector space $\mathcal{H}_{2,2}$ for evaluation order two consists of six vector monomials:

$$\mathcal{H}_{2,2} = \left\{ \begin{pmatrix} z_1^2 \\ 0 \end{pmatrix}, \begin{pmatrix} z_1 z_2 \\ 0 \end{pmatrix}, \begin{pmatrix} z_2^2 \\ 0 \end{pmatrix}, \begin{pmatrix} 0 \\ z_1^2 \end{pmatrix}, \begin{pmatrix} 0 \\ z_1 z_2 \end{pmatrix}, \begin{pmatrix} 0 \\ z_2^2 \end{pmatrix} \right\} .$$

The quadratic terms of the Taylor series evaluation of a two-dimensional vector space (4.6) can be written with the twofold indexing of the coefficients as follows:

$$\begin{aligned} \boldsymbol{P}_2(\boldsymbol{z}) = \; & a_{2,1,1} \begin{pmatrix} z_1^2 \\ 0 \end{pmatrix} + a_{2,1,2} \begin{pmatrix} z_1 z_2 \\ 0 \end{pmatrix} + a_{2,1,3} \begin{pmatrix} z_2^2 \\ 0 \end{pmatrix} + \\ & a_{2,2,1} \begin{pmatrix} 0 \\ z_1^2 \end{pmatrix} + a_{2,2,2} \begin{pmatrix} 0 \\ z_1 z_2 \end{pmatrix} + a_{2,2,3} \begin{pmatrix} 0 \\ z_2^2 \end{pmatrix} . \end{aligned}$$

By means of a Taylor series expansion the following considerations are structurally simplified because the problems can be expressed in matrix notation. Table 4.2 shows the dimension $n \cdot \hat{l}_k$ of the vector spaces in dependence of the system dimension n and the evaluation order k. Even for low dimensions and evaluation orders very many coefficients have to be considered.

Table 4.2: Dimensions of the vector spaces over the vectorial polynomials

	$k=1$	$k=2$	$k=3$	$k=4$	$k=5$
$n=1$	1	1	1	1	1
$n=2$	2	6	8	10	12
$n=3$	3	18	30	45	63
$n=4$	4	40	80	140	224
$n=5$	5	75	175	350	630

The multivariate Taylor expansion just constitutes the starting point for the most relevant question: *is it possible to transform the flow of the vector field (4.1) by a nonlinear transformation into a simplified form with identical local topological properties?*

Vector fields which can be mapped onto each other by such transformations are called *equivalent*, TROGER AND STEINDL (1991). They locally exhibit the same qualitative behavior and are discussed in the following.

Near Identity Transformations

Because the linear part of (4.2) should remain unchainged under the transformation formal coordinate transformations into linear identical variables with an additional nonlinearity $\boldsymbol{h}_k$ are considered, LICHTENBERG (1992):

$$\boldsymbol{\varphi} : \boldsymbol{z} \rightarrow \boldsymbol{x} + \boldsymbol{h}_k(\boldsymbol{x}) \; , \; k = 2, \ldots, s \tag{4.12}$$

For the determination of the transformed map equations $\boldsymbol{\varphi}^{-1} \circ \boldsymbol{p} \circ \boldsymbol{\varphi}$ an inversion of the power series $\boldsymbol{\varphi}(\boldsymbol{x})$ is necessary. The local existence of the inverse $\boldsymbol{\varphi}^{-1}$ is ensured because $\boldsymbol{\varphi}$ is nearly the identity map. A Taylor series expansion of the inverse yields

$$\boldsymbol{\varphi}^{-1}(\boldsymbol{x}) = \boldsymbol{x} - \boldsymbol{h}_k(\boldsymbol{x}) + \overline{\boldsymbol{h}}_{k+1}(\boldsymbol{x}) + \ldots + \overline{\boldsymbol{h}}_s(\boldsymbol{x}) \; . \tag{4.13}$$

Hence, the nonlinear term of lowest order is simply given by $-\boldsymbol{h}_k(\boldsymbol{x})$ as is easily confirmed by the composition of the maps $\boldsymbol{\varphi} \circ \boldsymbol{\varphi}^{-1}(\boldsymbol{x})$

$$\boldsymbol{\varphi}^{-1}(\boldsymbol{\varphi}(\boldsymbol{x})) = \boldsymbol{x} + \boldsymbol{h}_k(\boldsymbol{x}) - \boldsymbol{h}_k(\boldsymbol{x} + \boldsymbol{h}_k(\boldsymbol{x})) + \mathcal{O}(\|\boldsymbol{x}^{k+1}\|) = \boldsymbol{x} + \mathcal{O}(\|\boldsymbol{x}^{k+1}\|) \; . \tag{4.14}$$

The higher-order terms of the inverse transformation can be determined by calculating $\boldsymbol{\varphi}^{-1} \circ \boldsymbol{\varphi}$ using the general ansatz (4.13) and determining the sought transformation polynomials $\overline{\boldsymbol{h}}_{k+1}(\boldsymbol{x}), \ldots, \overline{\boldsymbol{h}}_s(\boldsymbol{x})$ such that all coefficients of the result are identically zero up to the maximum degree s:

$$\boldsymbol{\varphi}^{-1}(\boldsymbol{\varphi}(\boldsymbol{x})) \stackrel{!}{=} \boldsymbol{x} + \mathcal{O}(\|\boldsymbol{x}^{s+1}\|) \; . \tag{4.15}$$

One starts with nonlinear vector polynomials of second order $\boldsymbol{h}_2$ contained in the vector space $\mathcal{H}_{2,2}$:

$$\boldsymbol{z} \rightarrow \boldsymbol{x} + \boldsymbol{h}_2\,(\boldsymbol{x})\,. \tag{4.16}$$

This yields for the transformed right hand sides

$$\begin{aligned} \varphi^{-1} \circ \boldsymbol{p} \circ \varphi &= \varphi^{-1}\left[\,\boldsymbol{p}\,(\,\varphi\,(\boldsymbol{x})\,)\,\right], \\ &= \varphi^{-1}\left[\,\boldsymbol{p}\,(\boldsymbol{x} + \boldsymbol{h}_2(\boldsymbol{x})\,)\,\right], \\ &= \varphi^{-1}\left[\,\boldsymbol{J}\,(\boldsymbol{x} + \boldsymbol{h}_2(\boldsymbol{x})\,) + \boldsymbol{P}_2(\boldsymbol{x} + \boldsymbol{h}_2(\boldsymbol{x})\,) + \mathcal{O}\,(\|\boldsymbol{x}\|^3)\,\right]. \end{aligned} \tag{4.17}$$

Thereby, the vector-valued polynomial $\boldsymbol{P}_2\,(\boldsymbol{x}+\boldsymbol{h}_2(\boldsymbol{x}))$ can contain terms up to fourth order in $\boldsymbol{x}$ but for now all terms of order higher than the transformation order two are neglected:

$$\boldsymbol{P}_2\,(\boldsymbol{x} + \boldsymbol{h}_2\,(\boldsymbol{x})) = \boldsymbol{P}_2\,(\boldsymbol{x}) + \tilde{\boldsymbol{P}}_3\,(\boldsymbol{x}) + \tilde{\boldsymbol{P}}_4\,(\boldsymbol{x}) = \boldsymbol{P}_2\,(\boldsymbol{x}) + \mathcal{O}\,(\|\boldsymbol{x}\|^3)\,. \tag{4.18}$$

Therefore, using (4.17) with (4.13) and (4.18) the transformed system evaluated up to order two is given by

$$\varphi^{-1} \circ \boldsymbol{p} \circ \varphi = \boldsymbol{J}\,\boldsymbol{x} + \boldsymbol{J}\,\boldsymbol{h}_2(\boldsymbol{x}) + \boldsymbol{P}_2(\boldsymbol{x}) - \boldsymbol{h}_2\,(\boldsymbol{J}\,\boldsymbol{x}) + \mathcal{O}\,(\|\boldsymbol{x}\|^3)\,. \tag{4.19}$$

In the linear part it is identical to the original system (4.2). The simplest transformed system equations one would obtain if all nonlinear parts were identically zero:

$$\boldsymbol{J}\,\boldsymbol{h}_2\,(\boldsymbol{x}) + \boldsymbol{P}_2(\boldsymbol{x}) - \boldsymbol{h}_2(\boldsymbol{J}\,\boldsymbol{x}) \stackrel{!}{=} \boldsymbol{0}\,. \tag{4.20}$$

The question now reads: *is it possible to choose* $\boldsymbol{h}_2\,(\boldsymbol{x})$ *such that (4.20) is fulfilled?*

Eq. (4.20) can be represented as a linear system of equations in the space of vector polynomials of degree two because the vectorial polynomials of a specified order span a linear vector space.

The Linear Operator $\mathcal{L}$

For the following investigation the definition of an operator $\mathcal{L}$ is useful which is given for a general square matrix $\boldsymbol{A}$ by

$$\mathcal{L}_A\,(\boldsymbol{h}\,(\boldsymbol{x})) \equiv \boldsymbol{h}(\boldsymbol{A}\,\boldsymbol{x}) - \boldsymbol{A}\,\boldsymbol{h}(\boldsymbol{x})\,. \tag{4.21}$$

The operator $\mathcal{L}_A$ is linear and *endomorph* for every order k , i.e. it maps all vectors of a vector space $\mathcal{H}_{k,n}$ again onto vectors of the same space ($\mathcal{L}_A : \mathcal{H}_{k,n} \to \mathcal{H}_{k,n}$). It is therefore *isomorph* to linear maps which can be represented by square matrices.

Therefore, the operator $\mathcal{L}_J$ can be represented by an $(n \cdot \hat{l}_k \times n \cdot \hat{l}_k)$–matrix $\boldsymbol{L}_k$. The matrix $\boldsymbol{L}_k$ provides for the linear transformation $\boldsymbol{L}_k : \mathbb{R}^{n \cdot \hat{l}_k} \to \mathbb{R}^{n \cdot \hat{l}_k}$. This relation is shown in the following diagram:

$$\begin{array}{ccc} \boldsymbol{h}_k & \xrightarrow{\mathcal{L}_J} & \mathcal{L}_J\,(\boldsymbol{h}_k) \\ \Updownarrow & & \Updownarrow \\ \boldsymbol{\alpha}_k & \xrightarrow{L_k} & \boldsymbol{L}_k \cdot \boldsymbol{\alpha}_k \end{array}$$

The vertical arrows symbolize the equivalence of the representation of multivariate power series of order k as vector polynomials $\boldsymbol{h}_k$ on the one hand and as coefficient vectors $\boldsymbol{\alpha}_k$ on the other hand.

The representation of a linear operator $\mathcal{L}_J$ as a square matrix corresponds to the l–indexing notation (4.9). According to the definition of linear maps the image of the l–th basis vector appears in the l–th column of the representing matrix $\boldsymbol{L}$.

The calculation of a matrix representation of $\mathcal{L}_J$ will be performed exemplarily for the Jacobian

$$\boldsymbol{J} = \begin{pmatrix} 1 & 1 \\ 0 & 1 \end{pmatrix} .$$

First, the images of all the basis vectors of $\mathcal{H}_{2,2}$ are calculated:

$$\mathcal{L}_J \begin{pmatrix} x_1^2 \\ 0 \end{pmatrix} = \begin{pmatrix} (x_1+x_2)^2 \\ 0 \end{pmatrix} - \begin{pmatrix} x_1^2 \\ 0 \end{pmatrix} = \begin{pmatrix} 2x_1x_2 + x_2^2 \\ 0 \end{pmatrix} ,$$

$$\mathcal{L}_J \begin{pmatrix} x_1x_2 \\ 0 \end{pmatrix} = \begin{pmatrix} (x_1+x_2)\,x_2 \\ 0 \end{pmatrix} - \begin{pmatrix} x_1x_2 \\ 0 \end{pmatrix} = \begin{pmatrix} x_2^2 \\ 0 \end{pmatrix} ,$$

$$\mathcal{L}_J \begin{pmatrix} x_2^2 \\ 0 \end{pmatrix} = \begin{pmatrix} x_2^2 \\ 0 \end{pmatrix} - \begin{pmatrix} x_2^2 \\ 0 \end{pmatrix} = \begin{pmatrix} 0 \\ 0 \end{pmatrix} ,$$

$$\mathcal{L}_J \begin{pmatrix} 0 \\ x_1^2 \end{pmatrix} = \begin{pmatrix} 0 \\ (x_1+x_2)^2 \end{pmatrix} - \begin{pmatrix} x_1^2 \\ x_1^2 \end{pmatrix} = \begin{pmatrix} -x_1^2 \\ 2x_1x_2 + x_2^2 \end{pmatrix} ,$$

$$\mathcal{L}_J \begin{pmatrix} 0 \\ x_1 x_2 \end{pmatrix} = \begin{pmatrix} 0 \\ (x_1 + x_2)\, x_2 \end{pmatrix} - \begin{pmatrix} x_1 x_2 \\ x_1 x_2 \end{pmatrix} = \begin{pmatrix} -x_1 x_2 \\ x_2^2 \end{pmatrix},$$

$$\mathcal{L}_J \begin{pmatrix} 0 \\ x_2^2 \end{pmatrix} = \begin{pmatrix} 0 \\ x_2^2 \end{pmatrix} - \begin{pmatrix} x_2^2 \\ x_2^2 \end{pmatrix} = \begin{pmatrix} -x_2^2 \\ 0 \end{pmatrix}.$$

Two different points of view arise if the action of the operator $\mathcal{L}$ of order two applied to the first basis vector is considered:

$$\begin{array}{ccc} \begin{pmatrix} x_1^2 \\ 0 \end{pmatrix} & \xrightarrow{\mathcal{L}_J} & \begin{pmatrix} 2x_1x_2 + x_2^2 \\ 0 \end{pmatrix} \\ \Updownarrow & & \Updownarrow \\ (\,1\,0\,0\,0\,0\,0\,)^T & \xrightarrow{L_2} & (\,0\,2\,1\,0\,0\,0\,)^T \end{array}$$

Therefore, the matrix representation of the operator $\mathcal{L}$ is given by the transformation of the images of the basis vectors to coefficient vectors which constitute the columns of the matrix $\boldsymbol{L}_2$:

$$\boldsymbol{L}_2 = \begin{pmatrix} 0 & 0 & 0 & -1 & 0 & 0 \\ 2 & 0 & 0 & 0 & -1 & 0 \\ 1 & 1 & 0 & 0 & 0 & -1 \\ 0 & 0 & 0 & 0 & 0 & 0 \\ 0 & 0 & 0 & 2 & 0 & 0 \\ 0 & 0 & 0 & 1 & 1 & 0 \end{pmatrix}.$$

The inhomogenious system of equations

$$\mathcal{L}_J\,(\boldsymbol{h}_k\,(\boldsymbol{x})) = \tilde{\boldsymbol{P}}_k(\boldsymbol{x}) \tag{4.22}$$

is called *homological equation* of order k, ARNOLD (1987). Equation (4.20) is therefore a homological equation of order two.

A reformulation of the question now reads: *does the image of the operator $\mathcal{L}_J$ in the space of second-order vector polynomials $\mathcal{H}_{2,2}$ contain the known vector field $\boldsymbol{P}_2\,(\boldsymbol{x})$?*

In general the homological equation (4.22) can for each order k be explicitly written as an inhomogenious linear system of equations:

$$\boldsymbol{L}_k\,\boldsymbol{\xi}_k = \boldsymbol{\alpha}_k\,. \tag{4.23}$$

Equation (4.23) has to be solved with respect to the coefficient vector $\boldsymbol{\xi}_k$ of the nonlinear transformation equations (4.12). Its components determine the transformation

$$\xi_{k,l} \equiv h_{k,i,j} \ , \ l = 1, \ldots, n \cdot \hat{l}_k \ . \tag{4.24}$$

The solvability of (4.23) is limited by the *Fredholm alternative*: for square matrices $\boldsymbol{L}_k$ one of the following alternatives is valid:

- The corresponding homogenious system of equations has only the trivial solution. Then the inhomogenious system (4.23) has a unique solution for arbitrary right hand sides $\boldsymbol{\alpha}_k$:

$$\boldsymbol{\xi}_k = \boldsymbol{L}_k^{-1} \boldsymbol{\alpha}_k \ . \tag{4.25}$$

- The homogenious system has a non-trivial solution. Then the inhomogenious system has a solution if and only if $\boldsymbol{\alpha}_k$ is orthogonal to each

$$\boldsymbol{\gamma}_k \in \ker(\boldsymbol{L}_k^T) \ . \tag{4.26}$$

For the first case one obtains a coefficient vector $\boldsymbol{\xi}_k$ of the transformation polynomials $\boldsymbol{h}_k$ by the unique solution (4.25) of Eq. (4.23). For the calculation of the transformation equations the inversion of the matrix $\boldsymbol{L}_k$ is necessary.

Using the representation $\boldsymbol{L}_k$ of the linear operator $\mathcal{L}$ the minimum order of resonance $\hat{k}$ can be determined. It is defined as the lowest order k for which $\boldsymbol{L}_k$ becomes singular, i.e. for which the second Fredholm alternative holds. If the order of resonance is known the coordinate transformation can be performed recursively with vector polynomials $\boldsymbol{h}_k$ starting with the transformation order $k = 2$ up to the order of resonance $k = \hat{k}$.

With the assumption that all terms of orders less than k are already eliminated by near identity transformations the transformed right hand sides for an arbitrary transformation order k follow as for transformation order $k = 2$ (Eq. 4.19).

The solution of the homological equation (4.22) of order k determines the transformation coefficients $h_{k,i,j}$. Thereby, in the evaluations (4.2) and (4.18) all terms up to order $\hat{k}$ have to be considered for every near identity transformation of order $k \leq \hat{k}$ because terms of higher order are affected by transformations of lower order.

Resonant Vector Polynomials

If the second case of the Fredholm alternative is valid the vector space can be decomposed into a subspace which is spanned by the image of the operator and its complement:

$$\mathcal{H}_{k,n} \equiv \text{im}\,(\mathcal{L}_J) \oplus \text{im}\,(\mathcal{L}_J)^{\perp} . \tag{4.27}$$

The generating system of the image is given by the column vectors of $\boldsymbol{L}$. The following definition generally holds:

The normal form of a nonlinear vector field is contained in the complement of the image of the linear operator $\mathcal{L}_J$.

The choice of the basis vectors of the complement $\boldsymbol{g}_{\hat{k},i} \in \text{im}\,(\mathcal{L}_J)^{\perp}$ can be done taking advantage of the Fredholm alternative. Therefore, one searches for vectors which are contained in the kernel of the transposed matrix $\ker\,(\boldsymbol{L}_{\hat{k}}^T)$.

By rewriting the transformed system (4.19) for the transformation order $\hat{k}$ one gets

$$\boldsymbol{J}\,\boldsymbol{x} - \mathcal{L}_J(\boldsymbol{h}_{\hat{k}}(\boldsymbol{x})) + \boldsymbol{P}_{\hat{k}}(\boldsymbol{x}) + \mathcal{O}\left(\|\boldsymbol{x}\|^{\hat{k}+1}\right) = \boldsymbol{J}\,\boldsymbol{x} + \boldsymbol{G}_{\hat{k}}(\boldsymbol{x}) + \mathcal{O}\left(\|\boldsymbol{x}\|^{\hat{k}+1}\right) . \tag{4.28}$$

The terms $\boldsymbol{G}_{\hat{k}}$ which can not be eliminated determine the nonlinear normal form of order $\hat{k}$. They are solutions of the following system of equations:

$$\mathcal{L}_J(\boldsymbol{h}_{\hat{k}}(\boldsymbol{x})) + \boldsymbol{G}_{\hat{k}}(\boldsymbol{x}) = \boldsymbol{P}_{\hat{k}}(\boldsymbol{x}) . \tag{4.29}$$

The terms of $\boldsymbol{G}_{\hat{k}}$ are called *resonant vector polynomials* of order $\hat{k}$. Their number r is equal to the defect of the representing matrix $\boldsymbol{L}_{\hat{k}}$. The defect of the $(n \cdot \hat{l}_{\hat{k}} \times n \cdot \hat{l}_{\hat{k}})$–matrix $\boldsymbol{L}_{\hat{k}}$ is given by the difference of the row-dimension and the rank:

$$r = n \cdot \hat{l}_{\hat{k}} - \text{rank}\,[\boldsymbol{L}_{\hat{k}}] = \dim\,[\text{im}(\mathcal{L}_J)^{\perp}] . \tag{4.30}$$

A nonlinear normal form consists of the basis vectors $\boldsymbol{g}_{\hat{k},i}$ of the complement $\text{im}(\mathcal{L}_J)^{\perp}$ and their coefficients $d_{\hat{k},i}$:

$$\boldsymbol{G}_{\hat{k}}(\boldsymbol{x}) = \sum_{i=1}^{r} d_{\hat{k},i} \cdot \boldsymbol{g}_{\hat{k},i} . \tag{4.31}$$

Together with the coefficients $d_{\hat{k},i}$ the terms of order $\hat{k}$ of the normal form of the critical subsystem (4.1) are determined. If (4.29) is interpreted as a linear system of equations it is easy to see that it is under-determined. On the right hand side of the inhomogenious system of equations the coefficient vector $\boldsymbol{\alpha}_{\hat{k}}$ of the critical subsystem appears. The new solution vector has dimension $n \cdot \hat{l}_{\hat{k}} + r$. Therefore the system of equations is given by the matrix $\boldsymbol{L}_{\hat{k}}^{+}$ which is extended by the resonant vectors:

$$\underbrace{\left(L_{\hat{k}} \mid g_{\hat{k},1} \cdots g_{\hat{k},r} \right)}_{L^+_{\hat{k}}} \cdot \underbrace{\begin{pmatrix} h_{\hat{k},1,1} \\ h_{\hat{k},1,2} \\ \cdots \\ h_{\hat{k},n,l_{\hat{k}}} \\ d_{\hat{k},1} \\ d_{\hat{k},2} \\ \cdots \\ d_{\hat{k},r} \end{pmatrix}}_{\xi^+_{\hat{k}}} = \alpha_{\hat{k}} \ . \tag{4.32}$$

The choice of a basis for the normal form will be exemplarily carried out with the already known matrix representation L_2 (see page 31). The matrix L_2 is singular, the rows are linearly dependent, the fourth row is even vanishing. It follows that there is no transformation which eliminates the vector monomial at the fourth element of the basis. This resonance is a basis vector of the normal form and can be written in coordinates as

$$g_{2,1} = \begin{pmatrix} 0 \\ x_1^2 \end{pmatrix}$$

In terms of the Fredholm alternative this means that $(\ 0\ 0\ 0\ 1\ 0\ 0\ )^T$ is a left eigenvector of the matrix L with respect to the eigenvalue zero which is the same as a right eigenvector of the transposed matrix L^T with respect to the eigenvalue zero. The rank of the matrix L_2 is four, therefore there must be a second linearly independent vector which is contained in the complement. A possible second vector is $(\ 1\ 0\ 0\ 0\ \frac{1}{2}\ 0\ )^T$, the corresponding vector polynomial follows to

$$g_{2,2} = \begin{pmatrix} x_1^2 \\ \frac{1}{2} x_1 x_2 \end{pmatrix} .$$

The choice of this second vector is not unique. There arise many other possibilities from the combination of image vectors and vectors of the complement. The conclusion is: *the basis of a normal form is not unique.*

If one calculates apart from the minimum resonance order further terms of higher order these may depend on free parameters which arise from non-unique near identity transformations of lower orders of resonance, PONCE ET AL. (1991).

Diagonal Linear Part

If the Jacobian J is diagonal with the vector of the eigenvalues $\boldsymbol{\lambda}$ this is true also for the representing matrix L of the linear operator $\mathcal{L}_J$ and it follows for their eigenvalues, ARNOLD (1987):

$$\Lambda_{k,l} = \boldsymbol{\lambda}^{\boldsymbol{m}(k,j)} - \lambda_i \,. \tag{4.33}$$

The notation introduced above indicates the following:

$$\boldsymbol{\lambda}^{\boldsymbol{m}} = \prod_{i=1}^{n} \lambda_i^{m_i} = \lambda_1^{m_1} \cdots \lambda_n^{m_n} \,. \tag{4.34}$$

Here, the index l is defined by (4.9). The eigenvalue $\Lambda_{k,l}$ represents — geometrically interpreted — the factor of stretching of the vector monomial with the exponent tupel $\boldsymbol{m}(k,j)$ with respect to the component i under the action of the operator.

The *resonance condition* (4.33) is also valid for non-diagonal linear parts $\boldsymbol{J}$: if

$$\lambda_i = \boldsymbol{\lambda}^{\boldsymbol{m}(k,j)} \,, i = 1, \ldots, n \,,\; j = 1, \ldots, \hat{l}_k, \tag{4.35}$$

is true for the eigenvalues of the linear part $\boldsymbol{J}$ and an exponent tupel $\boldsymbol{m}(k,j)$, then the operator $\mathcal{L}_J$ is singular[8]. The order k of the exponent tupel $\boldsymbol{m}$ is called *order of resonance*. The so-called *resonant monomials* constitute (as eigenvectors to every zero eigenvalue of $\boldsymbol{L}$) the uniquely given normal form

$$\boldsymbol{G}_{\hat{k}} = \sum_{\Lambda_{\hat{k},l\,(i,j)}=0} d_{\hat{k},l(i,j)} \cdot \boldsymbol{x}^{\boldsymbol{m}(\hat{k},j)} \cdot \boldsymbol{e}_i \,. \tag{4.36}$$

In Table 4.3 all cases are summarized which can occur for near identity transformations with nonlinear terms of order k:

Table 4.3: Possible cases for normal forms

	$\boldsymbol{J}$ diagonal	$\boldsymbol{J}$ non-diagonal
Resonant eigenvalues	Unique normal form consisting of resonant monomials	Choice of a normal form in the complement of the image of $\mathcal{L}$
Non-resonant eigenvalues	Reduction to the linear terms by the transformation coefficients: $h_{k,i,j} = \frac{a_{k,i,j}}{\Lambda_{k,l}}$	Transformation coefficients given uniquely by the inverse of $\mathcal{L}$

[8]Consequently there are also zero eigenvalues of the representing matrix $\boldsymbol{L}_k$.

The most difficult case of a resonance with non-diagonal linear part may be explicitly treated again.

Considering the example it has become obvious that the normal form is in general not unique. The calculation of a normal form therefore must contain four steps:

1. Calculation of the representation $\boldsymbol{L}_{\hat{k}}$ of the linear operator $\mathcal{L}$,
2. Choice of resonant vector polynomials $\boldsymbol{G}_{\hat{k}}$,
3. Near identity transformation of resonance order (4.12),
4. Determination of the coefficients by solving Eq. (4.32).

4.2 Classification

If the critical subsystem is transformed to its nonlinear normal form a classification of the bifurcation scenario is possible. In order to account for the influence of the control parameter μ the normal form transformation has to be performed for the extended system (GUCKENHEIMER AND HOLMES (1986))

$$\boldsymbol{z}_{N+1} = \boldsymbol{p}(\boldsymbol{z}_N, \mu) \, , \tag{4.37}$$

$$\mu_{N+1} = \mu_N \, . \tag{4.38}$$

The simplified system is then given by

$$\boldsymbol{x}_{N+1} = \tilde{\boldsymbol{p}}(\boldsymbol{x}_N, \mu) \tag{4.39}$$

which is the unfolded normal form of the critical subsystem. The trivial equation (4.38) remains unchanged because it does not have any nonlinear part.

In the following, the simple one-dimensional non-trivial bifurcation equations are just shortly presented while the more complicated situation for the Hopf bifurcation with its two-dimensional non-trivial critical subsystem is discussed in more detail.

One-Dimensional Bifurcations Systems

The unfolded nonlinear normal form for the saddle-node bifurcation is given by

$$x_{N+1} = x_N + d \cdot \mu + a \cdot x_N^2 \tag{4.40}$$

and for the transcritical bifurcation by

$$x_{N+1} = x_N + d \cdot \mu x_N + a \cdot x_N^2 \,. \tag{4.41}$$

For $d > 0$ and $a < 0$ these bifurcation types are illustrated in Fig. 4.1.

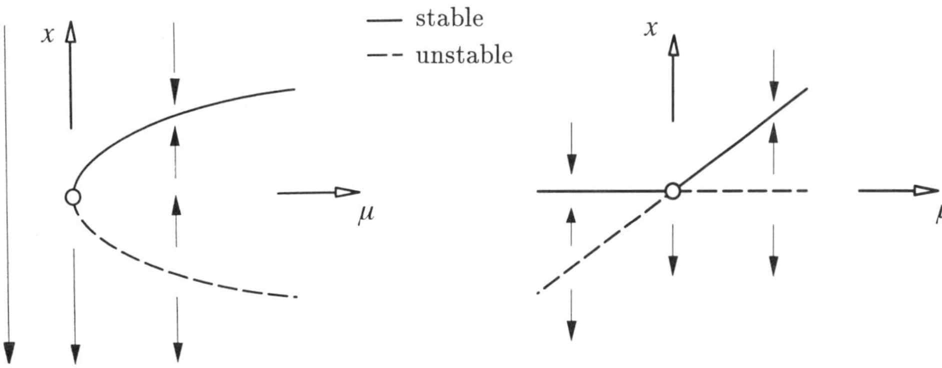

Figure 4.1: Saddle-node and transcritical bifurcation

The pitchfork bifurcation is a bifurcation with *symmetry breaking*; the considered periodic solution becomes unstable and two point-symmetric solutions with same periodicity branch off, Fig. 4.2.

The unfolded normal form is given by

$$x_{N+1} = x_N + d \cdot \mu x_N + a \cdot x_N^3 \,. \tag{4.42}$$

A flip bifurcation, which has the unfolded normal form

$$x_{N+1} = -x_N + \tilde{d} \cdot \mu x_N + \tilde{a} \cdot x_N^3 \,, \tag{4.43}$$

causes a so-called *period doubling*; the periodic solution of period T gets unstable and a new periodic solution of period $2T$ branches off the bifurcation point.

By considering the time-two map it becomes clear that the flip bifurcation can be traced back to a pitchfork bifurcation:

$$x_{N+2} = x_N \underbrace{-2\tilde{d}}\cdot \mu x_N \underbrace{-2\tilde{a}}\cdot x_N^3 \,, \tag{4.44}$$

$$x_{N+2} = x_N + d \cdot \mu x_N + a \cdot x_N^3 \,. \tag{4.45}$$

Therefore, the bifurcation diagrams are analogous to those shown in Fig. 4.2 except that the two non-trivial equilibrium branches belong to one single P–2 solution which intersects the surface of section alternately through the upper and the lower branch.

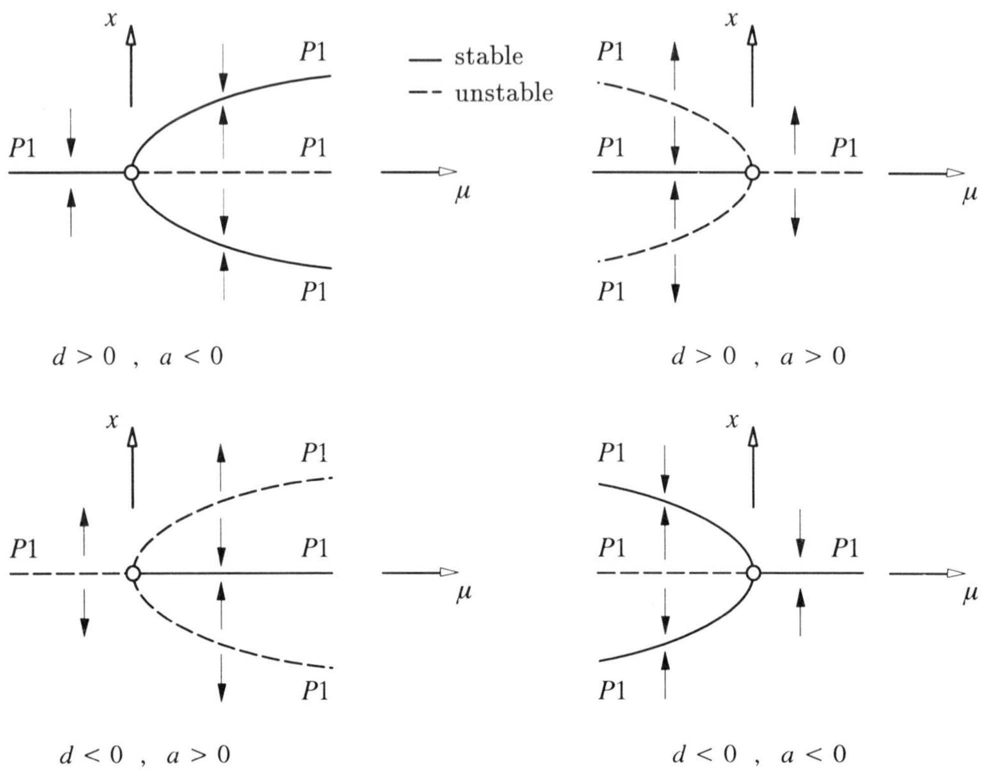

Figure 4.2: Pitchfork bifurcation diagrams

There are two different cases that can be distinguished for the pitchfork and the flip bifurcation according to the sign of the so-called *stability parameter* a: for a stable, called *supercritical* bifurcation the solutions branching off from the critical point are stable ($a < 0$) while those are unstable for an unstable, called *subcritical* bifurcation ($a > 0$).

This result is of major practical importance because after a subcritical bifurcation took place there is no stable equilibrium position in the neighborhood of the periodic reference trajectory. Therefore, the system will settle down to a stable equilibrium position which locally does not have anything to do with the reference solution. In this sense the saddle-node bifurcation is principally subcritical bifurcation.

Hopf Bifurcations

For a discrete Hopf bifurcation the Jacobian $\boldsymbol{J}$ can be described in terms of the argument ϑ of a complex eigenvalue:

$$\boldsymbol{J} = \begin{pmatrix} \cos\vartheta & -\sin\vartheta \\ \sin\vartheta & \cos\vartheta \end{pmatrix} . \tag{4.46}$$

With the eigenvalues $e^{\pm j\vartheta}$ the resonance condition (4.35) can be stated as

$$e^{\pm j\vartheta} = e^{j\vartheta m_1}\, e^{-j\vartheta m_2} = e^{j\Delta m\vartheta} \tag{4.47}$$

where the difference of the exponents is given by $\Delta m = m_1 - m_2$. Hence, the following resonance condition holds:

$$\Delta m = \pm 1 . \tag{4.48}$$

Figure 4.3 shows the position of this resonances in the (m_1, m_2)–plane.

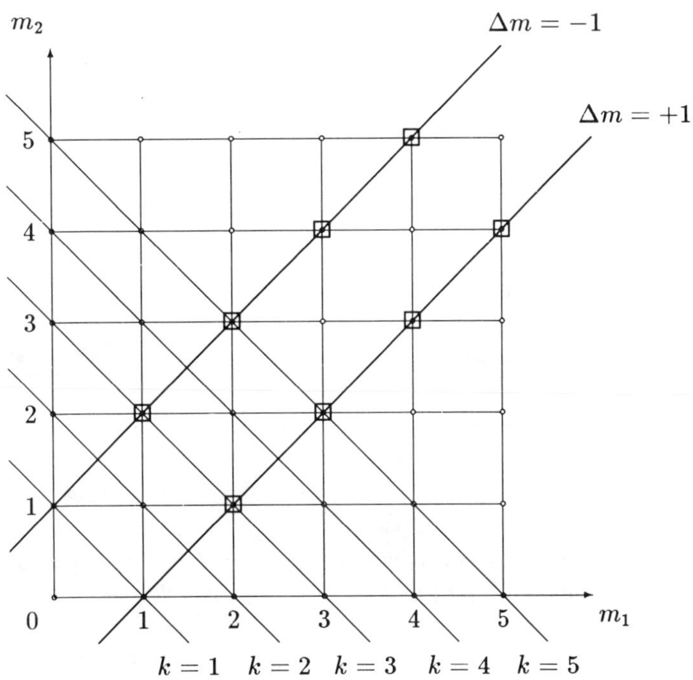

Figure 4.3: Resonances with conjugate-complex eigenvalues

It follows that no resonances with even orders k can appear while all odd orders are resonant. To every odd evaluation order there correspond exactly two resonances as marked in Fig. 4.3 by small boxes.

There exists a normal form basis which is particularly advantageous with respect to the stability analysis:

$$\boldsymbol{g}_{2i+1,1} = \begin{pmatrix} x_1(x_1^2+x_2^2)^i \\ x_2(x_1^2+x_2^2)^i \end{pmatrix} , \quad \boldsymbol{g}_{2i+1,2} = \begin{pmatrix} -x_2(x_1^2+x_2^2)^i \\ x_1(x_1^2+x_2^2)^i \end{pmatrix} . \tag{4.49}$$

Using this basis the normal form in cartesian coordinates $[x_1, x_2]$ results in

$$\boldsymbol{x}_{N+1} = \boldsymbol{J} \cdot \boldsymbol{x}_N + \sum_{i=1}^{\infty} \left(d_{2i+1,1} \cdot \boldsymbol{g}_{2i+1,1} + d_{2i+1,2} \cdot \boldsymbol{g}_{2i+1,2} \right)_N . \tag{4.50}$$

A transformation to polar coordinates $[R, \theta]$ with

$$x_1 = R \cdot \cos\theta \quad , \quad x_2 = R \cdot \sin\theta$$

yields

$$R_{N+1} = R_N + \Omega_{3,1} \cdot R_N^3 + \mathcal{O}(R_N^5) , \tag{4.51}$$

$$\theta_{N+1} = \theta_N + \vartheta + \Omega_{3,2} \cdot R_N^2 + \mathcal{O}(R_N^4) . \tag{4.52}$$

Thereby, $\Omega_{3,1}$ and $\Omega_{3,2}$ follow from the coefficients of the resonant vector polynomials:

$$\begin{aligned} \Omega_{3,1} &= d_{3,1}\cos\vartheta + d_{3,2}\sin\vartheta , \\ \Omega_{3,2} &= d_{3,2}\cos\vartheta - d_{3,1}\sin\vartheta \end{aligned} \tag{4.53}$$

with $d_{3,1}$ and $d_{3,2}$ as given below:

$$\begin{aligned} d_{3,1} = {} & (-6a_{3,1,1}\cos\vartheta - 12a_{3,1,1}\cos^2\vartheta + 2{a_{2,1,1}}^2\cos\vartheta + 2a_{3,2,2}\cos\vartheta + 4{a_{2,2,3}}^2\cos^3\vartheta \\ & -12a_{3,2,4}\cos^2\vartheta - 4{a_{2,1,3}}^2\cos^3\vartheta - 4{a_{2,2,1}}^2\cos^3\vartheta - 4a_{3,2,2}\cos^2\vartheta + 6a_{2,1,1}a_{2,1,3} \\ & -4a_{3,1,3}\cos^2\vartheta + a_{2,1,2}a_{2,2,1}\cos\vartheta + 4{a_{2,1,1}}^2\cos^3\vartheta + 6a_{3,2,4}\cos\vartheta + a_{2,1,3}a_{2,2,2}\cos\vartheta \\ & -8{a_{2,1,1}}^2\cos^2\vartheta + 3a_{2,1,1}a_{2,2,2} + 2{a_{2,2,3}}^2\cos\vartheta + 6a_{2,2,3}a_{2,2,1} - 8{a_{2,2,3}}^2\cos^2\vartheta \\ & +2a_{3,1,3}\cos\vartheta + 4{a_{2,1,3}}^2\cos\vartheta + 3a_{2,1,2}a_{2,2,3} + 4{a_{2,2,1}}^2\cos\vartheta + 4a_{2,2,3}a_{2,1,2}\cos^3\vartheta \\ & -a_{2,1,2}a_{2,2,1} + 4a_{2,1,1}a_{2,1,2}\cos\vartheta\sin\vartheta - 4a_{2,2,2}a_{2,2,3}\sin\vartheta\cos\vartheta + 4a_{2,1,2}a_{2,1,3}\sin\vartheta\cos\vartheta \\ & -3a_{2,1,1}a_{2,2,2}\cos\vartheta - 6a_{2,1,1}a_{2,2,1}\sin\vartheta - 8a_{2,1,1}a_{2,2,1}\sin\vartheta\cos\vartheta - a_{2,1,3}a_{2,2,2} \\ & +6a_{2,2,3}a_{2,1,3}\sin\vartheta - 4a_{2,1,2}a_{2,2,1}\cos^2\vartheta + 4a_{2,2,1}a_{2,2,2}\sin\vartheta\cos^2\vartheta - 3a_{2,1,2}a_{2,2,3}\cos\vartheta \\ & +4a_{2,1,1}a_{2,2,2}\cos^3\vartheta + 2a_{2,1,3}a_{2,1,1}\cos\vartheta - 3a_{2,2,1}a_{2,2,2}\sin\vartheta + 4a_{2,2,3}a_{2,2,2}\sin\vartheta\cos^2\vartheta \\ & -4a_{2,1,3}a_{2,2,2}\cos^2\vartheta + 3a_{2,1,2}a_{2,1,3}\sin\vartheta + {a_{2,1,2}}^2\cos\vartheta - 4a_{2,1,2}a_{2,1,3}\sin\vartheta\cos^2\vartheta \\ & +2a_{3,2,2} + 6a_{3,2,4} + 6a_{3,1,1} + 2a_{3,1,3} + 4a_{2,2,1}a_{2,1,2}\cos^3\vartheta - 3a_{2,2,2}a_{2,2,3}\sin\vartheta \\ & +8a_{2,1,1}a_{2,2,1}\sin\vartheta\cos^2\vartheta - {a_{2,1,2}}^2 - 4a_{2,1,1}a_{2,1,2}\sin\vartheta\cos^2\vartheta - 4a_{2,1,1}a_{2,2,2}\cos^2\vartheta \\ & -8a_{2,2,3}a_{2,1,3}\sin\vartheta\cos^2\vartheta + 2{a_{2,2,3}}^2 + 4a_{2,1,3}a_{2,2,2}\cos^3\vartheta - 8a_{2,2,1}a_{2,2,3}\cos^2\vartheta + 2{a_{2,1,1}}^2 \\ & +2a_{2,2,1}a_{2,2,3}\cos\vartheta - 8a_{2,1,3}a_{2,1,1}\cos^2\vartheta + 8a_{2,2,3}a_{2,1,3}\sin\vartheta\cos\vartheta + 3a_{2,1,1}a_{2,1,2}\sin\vartheta \\ & -4a_{2,1,2}a_{2,2,3}\cos^2\vartheta - 4a_{2,2,2}a_{2,2,1}\sin\vartheta\cos\vartheta - {a_{2,2,2}}^2 + {a_{2,2,2}}^2\cos\vartheta) \\ & /16(1-\cos\vartheta)(1+2\cos\vartheta) \end{aligned}$$

$$\begin{aligned}
d_{3,2} = {} & (4a_{2,2,3}{}^2\cos^3\vartheta - 4a_{2,1,3}{}^2\cos^3\vartheta - 4a_{2,2,1}{}^2\cos^3\vartheta + 2a_{2,1,1}a_{2,1,3} - 5a_{2,1,2}a_{2,2,1}\cos\vartheta \\
& + 4a_{2,1,1}{}^2\cos^3\vartheta - 5a_{2,1,3}a_{2,2,2}\cos\vartheta + 4a_{2,1,1}{}^2\cos^2\vartheta - 5a_{2,1,1}a_{2,2,2} + 2a_{2,2,3}a_{2,2,1} \\
& + 4a_{2,2,3}{}^2\cos^2\vartheta + 14a_{2,1,3}{}^2\cos\vartheta - 5a_{2,1,2}a_{2,2,3} + 14a_{2,2,1}{}^2\cos\vartheta + 4a_{2,2,3}a_{2,1,2}\cos^3\vartheta \\
& - a_{2,1,2}a_{2,2,1} - 9a_{2,1,1}a_{2,2,2}\cos\vartheta - 2a_{2,1,1}a_{2,2,1}\sin\vartheta - a_{2,1,3}a_{2,2,2} \\
& + 2a_{2,2,3}a_{2,1,3}\sin\vartheta + 4a_{2,2,1}a_{2,2,2}\sin\vartheta\cos^2\vartheta - 9a_{2,1,2}a_{2,2,3}\cos\vartheta + 4a_{2,1,1}a_{2,2,2}\cos^3\vartheta \\
& + 10a_{2,1,3}a_{2,1,1}\cos\vartheta - a_{2,2,1}a_{2,2,2}\sin\vartheta + 4a_{2,2,3}a_{2,2,2}\sin\vartheta\cos^2\vartheta + a_{2,1,2}a_{2,1,3}\sin\vartheta \\
& + a_{2,1,2}{}^2\cos\vartheta - 4a_{2,1,2}a_{2,1,3}\sin\vartheta\cos^2\vartheta + 4a_{2,2,1}a_{2,1,2}\cos^3\vartheta - a_{2,2,2}a_{2,2,3}\sin\vartheta \\
& + 8a_{2,1,1}a_{2,2,1}\sin\vartheta\cos^2\vartheta + a_{2,1,2}{}^2 - 4\cos^2\vartheta a_{2,1,1} + a_{2,1,2}\sin\vartheta\cos^2\vartheta \\
& + 4a_{2,1,3}a_{2,2,2}\cos^3\vartheta + 8a_{2,2,1}a_{2,2,3}\cos^2\vartheta + 10a_{2,2,1}a_{2,2,3}\cos\vartheta + 8a_{2,1,3}a_{2,1,1}\cos^2\vartheta \\
& + a_{2,1,1}a_{2,1,2}\sin\vartheta + a_{2,2,2}{}^2 + a_{2,2,2}{}^2\cos\vartheta + 6a_{2,2,1}{}^2 - 8a_{2,2,3}a_{2,1,3}\sin\vartheta\cos^2\vartheta \\
& - 4a_{3,2,3}\sin\vartheta\cos\vartheta - 12a_{3,2,1}\sin\vartheta\cos\vartheta + 6a_{2,1,3}{}^2 + a_{2,1,1}a_{2,1,2}\sin\vartheta + a_{2,2,2}{}^2 \\
& + 4a_{3,1,2}\sin\vartheta\cos\vartheta + 12a_{3,1,4}\sin\vartheta\cos\vartheta + 6a_{3,1,4}\sin\vartheta - 6a_{3,2,1}\sin\vartheta + 4a_{2,1,3}{}^2\cos^2\vartheta \\
& + 4a_{2,2,1}{}^2\cos^2\vartheta - 2a_{3,2,3}\sin\vartheta + 2a_{3,1,2}\sin\vartheta - 8a_{2,2,3}a_{2,1,3}\sin\vartheta\cos^2\vartheta \\
& + 4a_{2,1,3}a_{2,2,2}\cos^3\vartheta + 8a_{2,2,1}a_{2,2,3}\cos^2\vartheta + 10a_{2,2,1}a_{2,2,3}\cos\vartheta + 8a_{2,1,3}a_{2,1,1}\cos^2\vartheta \\
& + a_{2,2,2}{}^2\cos\vartheta + 6a_{2,2,1}{}^2 - 4a_{3,2,3}\sin\vartheta\cos\vartheta - 12a_{3,2,1}\sin\vartheta\cos\vartheta \\
& + 6a_{2,1,3}{}^2 + 4a_{3,1,2}\sin\vartheta\cos\vartheta + 12a_{3,1,4}\sin\vartheta\cos\vartheta + 6a_{3,1,4}\sin\vartheta \\
& - 6a_{3,2,1}\sin\vartheta + 4a_{2,1,3}{}^2\cos^2\vartheta + 4a_{2,2,1}{}^2\cos^2\vartheta - 2a_{3,2,3}\sin\vartheta + 2a_{3,1,2}\sin\vartheta) \\
& / - 16\sin\vartheta(1 + 2\cos\vartheta)
\end{aligned}$$

Introducing an unfolding parameter μ vanishing in the bifurcation point results in the unfolded normal form in polar coordinates

$$R_{N+1} = R_N + \kappa_1 \cdot \mu R_N + \Omega_{3,1} \cdot R_N^3 + \mathcal{O}(\mu^2 R_N, \mu R_N^3, R_N^5)\,, \tag{4.54}$$

$$\theta_{N+1} = \theta_N + \vartheta + \kappa_2 \cdot \mu + \Omega_{3,2} \cdot R_N^2 + \mathcal{O}(\mu^2, \mu R_N^2, R_N^4)\,. \tag{4.55}$$

If a Hopf bifurcation for discrete systems takes place the periodic orbit evolves into a torus. The trajectories, depending on the frequency ratio, can be either closed or lay dense on the torus. In the first case a so-called *frequency locking* phenomenon is observed while in the second case the motion is called *quasi-periodic*.

Using (4.54) the investigation of the stability of the trajectories in the vicinity of the singularity is straightforward and can be summarized according to Fig. 4.4, WIGGINS (1990).

The stability parameter is given by $\Omega_{3,1}$ which indicates either a subcritical ($\Omega_{3,1} > 0$) or a supercritical bifurcation ($\Omega_{3,1} < 0$).

In the degenerate case $\Omega_{3,1} = 0$ the coefficient $\Omega_{5,1}$ has to be calculated. For real systems, however, this degeneracy is highly unlikely if the system does not possess special symmetry properties.

There exists another type of degeneracy which may occur for discrete Hopf bifurcations. Equation (4.47) must be satisfied for ϑ modulo 2π, i.e. with $n_1 \in \mathbb{Z}\backslash\{0\}$ follows also:

$$e^{\pm j\vartheta} = e^{\pm j(\vartheta + 2\pi n_1)} = e^{j\Delta m\vartheta}\,. \tag{4.56}$$

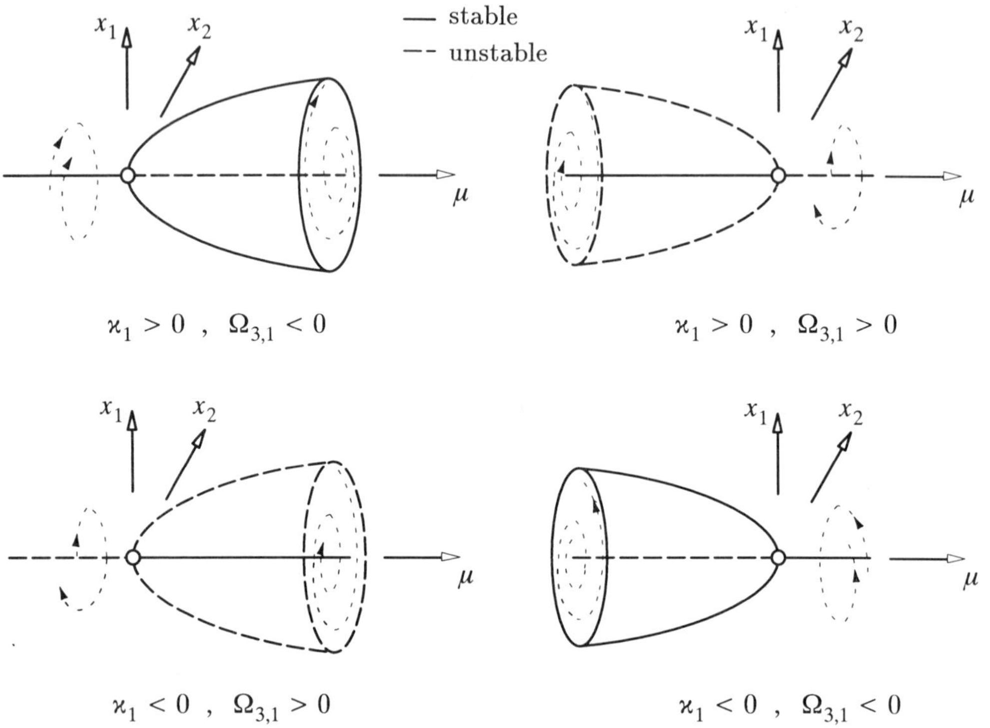

Figure 4.4: Hopf bifurcation diagrams

The difference of the exponents Δm is always integer-valued and defined by $(n_2 + 1)$ with $n_2 \in \mathbb{Z}\backslash\{0\}$:

$$\pm \Delta m \ = \ 1 + \frac{2\pi n_1}{\vartheta} \ = \ n_2 + 1 \ . \tag{4.57}$$

It follows

$$\frac{\vartheta}{2\pi} \ = \ \frac{n_1}{n_2} \ . \tag{4.58}$$

The conclusion is: if ϑ is in a rational relation to 2π further resonances can appear which correspond to exponent differences $\pm\Delta m$ as are given by (4.57). For a detailed discussion of the problems with respect to degenerate cases see ARNOLD (1987).

4.3 Computer Algebra Implementation

The flow diagram for the complete normal form transformation procedure is shown in Fig. 4.5, LICHTENBERG (1992).

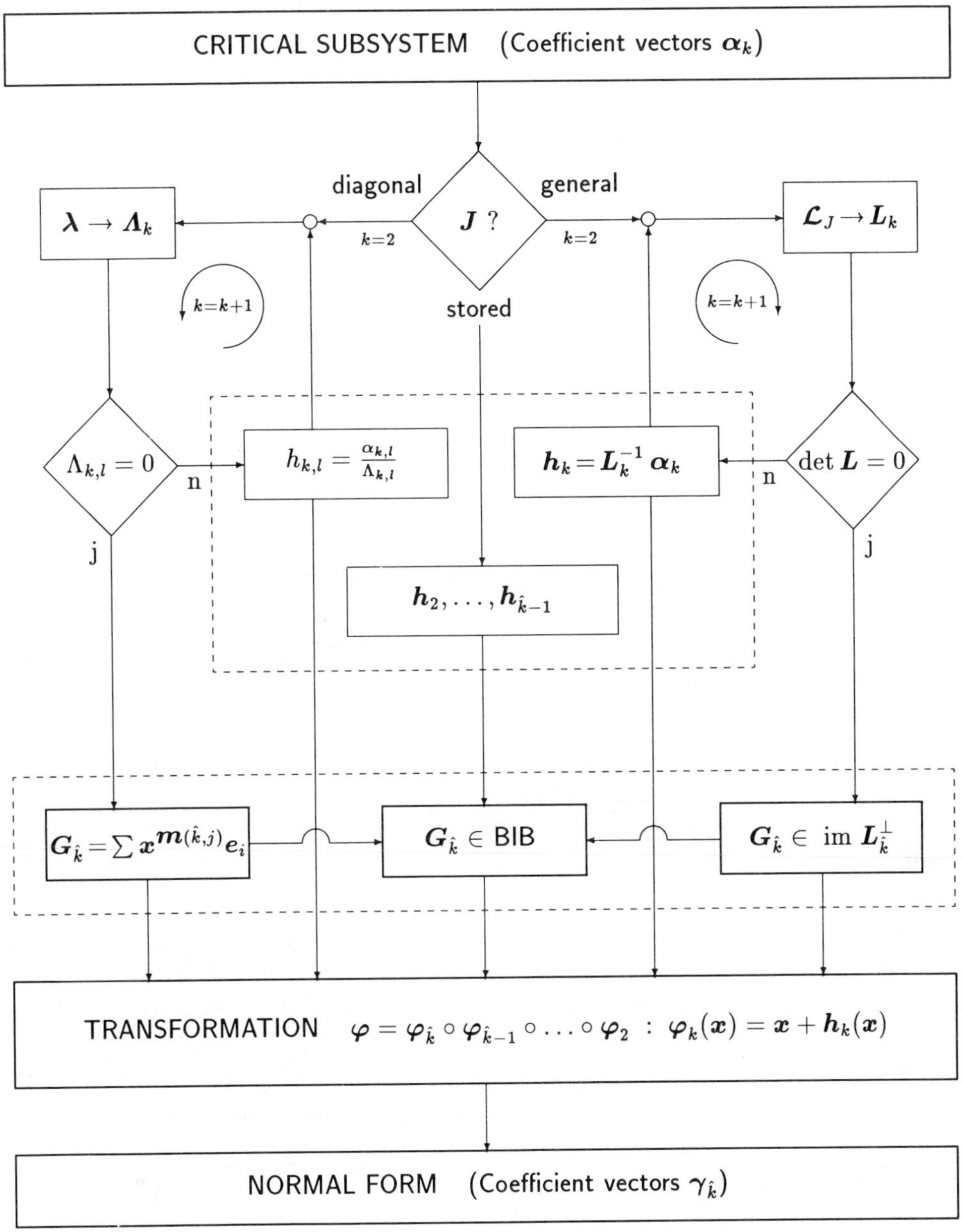

Figure 4.5: Flow diagram for the normal form transformation

In the top part of Fig. 4.5 those steps are arranged which only depend on the linear part. The computation of the matrix representation $\boldsymbol{L}$, its determinant, and (in the non-singular case) its inverse is shown in the right branch. For diagonal Jacobians the left branch holds.

In the non-resonant case the order k is increased by one and the loop is started again. This recursion is performed as long as no resonance is found or the maximum order is reached.

Each run of this loop produces either the transformation equations $\boldsymbol{h}_k$ or a basis of the normal form. In the non-diagonal case the choice of different bases is possible. For this a normal form library has been installed which provides normal form bases for different classes of linear parts. All bases for normal forms are collected in Fig. 4.5 in the lower dashed box.

All transformations with non-resonant orders less than the order of the normal form are independent of the choice of the basis. They can be saved in terms of the inverse matrix $\boldsymbol{L}_k^{-1}$ to avoid multiple calculations. The top dashed box contains all transformation coefficients.

After the determination of the resonances the transformed critical subsystem is computed up to the resonance order with parametrized representations of the transformation coefficients. The transformation coefficients of the non-resonant orders are inserted afterwards to obtain a critical subsystem which contains only nonlinear terms of the resonance order.

The determination of the coefficients of the normal form basis is the last step and is done by comparison of the coefficients. This is trivial in the case of a diagonal Jacobian. For a non-diagonal linear part the transformation in the resonance order leads to a linear system of equations (4.32) which can be solved by standard methods (for example with Maple by the *solve*-function).

In Table 4.4 the computation time and memory requirements for the calculation of the total parametrized normal form for continuous and discrete Hopf bifurcations are given for the main frame computer at the Technical University Hamburg-Harburg (CONVEX C 240).

Table 4.4: Computation times and memory requirements

Hopf normal form	Order	Computation time [s]	Memory space [Mbyte]
Continuous	3	21	3
Continuous	5	436	45
Discrete	3	1288	166
Discrete	5	3449	352

5 Computational Aspects

It is necessary to define interfaces between the symbol-manipulating and numerical routines which are in general valid for the concept introduced in Chapter 1. In particular, a general data type (`system`) for continuous as well as discrete systems has been introduced which contains the standard-forms of the right hand sides, the parameter vector, the state variables, etc. Program modules can be constructed such that both the argument and the result are of the type `system`. Then, for example, a transformation program changes the right hand sides and the state variables but leaves the parameter vector invariant.

5.1 Coefficient Representations and Basic Operations

As is explained in Chapter 4 the computation of a normal form is carried out in the space of vector polynomials of a certain dimension n and a finite evaluation order k. If a basis of this vector space is chosen a vector polynomial is uniquely determined by the coefficients of the multivariate Taylor expansion. Now the question arises: which operations have to be introduced for the elements of this vector space? The answer is: just the four operations *addition*, *multiplication*, *substitution*, and *partial differentiation* are necessary.

Evaluating the order s of the result of an operation in dependence of the orders of the arguments s_1, s_2 general relations can be stated as is given in Table 5.1.

Table 5.1: Order of the results of basic operations

Operation	Order s	Representation
Addition	$\max(s_1, s_2)$	$Q_1(\boldsymbol{x}) + Q_2(\boldsymbol{x})$
Multiplication	$s_1 + s_2$	$Q_1(\boldsymbol{x}) * Q_2(\boldsymbol{x})$
Substitution	$s_2 * s_1$	$Q_1(\boldsymbol{Q}_2(\boldsymbol{x}))$
Differentiation	$s_1 - 1$	$\partial Q_1(\boldsymbol{x})/\partial x_i$

The results of Table 5.1 are also valid for vectorial polynomials $(\boldsymbol{Q}_1, \boldsymbol{Q}_2)$ for which the operations are defined for each component. For the multiplication and substitution the order of the result exceeds the order of the operands. Such orders are not significant and can therefore be neglected. In order to avoid a tremendous overhead in the calculations a central demand for the realization of the concept is the strict avoidance of the computation of irrelevant orders.

Minimal Representation of Multivariate Power Series

The reduction of a general multivariate power series to a minimal form which still contains all information is the first step in the realization. The reduced representation of a multivariate power series does not contain the symbols for the state variables any more but exclusively the coefficients of the monomials with respect to a given basis. An example of the transformation of a polynomial into its coefficient representation is given below:

The multivariate polynomial

$$Q = 3x_1^2x_2 - 4x_1^2 + ax_1 + 9x_1^3 + 2x_2^2 - 3x_1x_2 - c$$

can be represented by its coefficients with respect to a basis of the polynomials in the variables $[x_1, x_2]$. If the monomials are arranged with ascending total order, and within each order with respect to the order of the single variables, one obtains

$$Q = -c + ax_1 - 4x_1^2 - 3x_1x_2 + 2x_2^2 + 9x_1^3 + 3x_1^2x_2 \ .$$

If the still missing monomials are completed by zero-coefficients up to total order three,

$$Q = -c + ax_1 + 0x_2 - 4x_1^2 - 3x_1x_2 + 2x_2^2 + 9x_1^3 + 3x_1^2x_2 + 0x_1x_2^2 + 0x_2^3 \ ,$$

then the coefficient representation ν_Q in the natural basis of the polynomials in the variables $[x_1, x_2]$ is uniquely given by

$$\nu_Q = [-c, a, 0, -4, -3, 2, 9, 3, 0, 0]$$

and vice versa.

For the identification of each coefficient a function must be given which relates each position $\jmath$ in the coefficient list to the corresponding n–tupel $\boldsymbol{m}$ of exponents.

Table 5.2 shows for a two-dimensional system this systematic relation up to the third evaluation order.

Table 5.2: Relation between list position and exponent tupels

$\jmath \leftrightarrow \boldsymbol{m}$	Monomial	Order
$1 \leftrightarrow [0,0]$	1	constant term
$2 \leftrightarrow [1,0]$	x_1	linear terms
$3 \leftrightarrow [0,1]$	x_2	
$4 \leftrightarrow [2,0]$	x_1^2	second order terms
$5 \leftrightarrow [1,1]$	x_1x_2	
$6 \leftrightarrow [0,2]$	x_2^2	
$7 \leftrightarrow [3,0]$	x_1^3	third order terms
$8 \leftrightarrow [2,1]$	$x_1^2x_2$	
$9 \leftrightarrow [1,2]$	$x_1x_2^2$	
$10 \leftrightarrow [0,3]$	x_2^3	

The exponent tupel $\boldsymbol{m}$ represents the monomial $\boldsymbol{x}^{\boldsymbol{m}}$ according to (4.3). Fast algorithms to relate a natural number $\jmath$ to the respective n-tupel $\boldsymbol{m}$ and vice versa are implemented in Maple V [9].

For the conversion of a multivariate function into its coefficient representation the evaluation basis with dimension and order must be given and the multivariate Taylor expansion must be performed. As a result of this considerations a new data type has been defined which is recognized by Maple and must be treated with special rules, LICHTENBERG (1992).

The elements of the coefficient representations again can be evaluated with respect to another basis. In principle, arbitrarily deep levels of evaluation are possible, e.g. for certain applications a further evaluation with respect to the system parameters is useful.

Computation Rules for Coefficient Representations

First, the basic operations 'multiplication' and 'addition' will be regarded with respect to the new data type. In the next examples the smallest non-degenerate evaluation of second order is considered. For the polynomials the coordinates $[x_1, x_2]$ are assumed and both the polynomials and the coefficients are indexed. For example, for the general coefficient representation ν_a of the polynomial Q_a one has:

$$\begin{aligned} Q_a &= a_1 + a_2 x_1 + a_3 x_2 + a_4 x_1^2 + a_5 x_1 x_2 + a_6 x_2^2 \,, \\ \nu_a &= [a_1, a_2, a_3, a_4, a_5, a_6] \,. \end{aligned}$$

For the addition of two coefficient representations just the list entries at the same position have to be added:

$$\begin{aligned} Q_a + Q_b &= (a_1 + b_1) + (a_2 + b_2) x_1 + (a_3 + b_3) x_2 + \\ &\quad (a_4 + b_4) x_1^2 + (a_5 + b_5) x_1 x_2 + (a_6 + b_6) x_2^2 \,, \\ \nu_a + \nu_b &= [a_1 + b_1, a_2 + b_2, a_3 + b_3, a_4 + b_4, a_5 + b_5, a_6 + b_6] \,. \end{aligned}$$

The procedure is trivial. The multiplication results in a more complex structure; sums of products of coefficients of the operands occur:

$$\begin{aligned} Q_a * Q_b &= (a_1 + a_2 x + a_3 y + a_4 x^2 + a_5 xy + a_6 y^2) * \\ &\quad (b_1 + b_2 x + b_3 y + b_4 x^2 + b_5 xy + b_6 y^2) \,, \\ \nu_a * \nu_b &= [a_1 b_1, a_2 b_1 + a_1 b_2, a_3 b_1 + a_1 b_3, a_4 b_1 + a_2 b_2 + a_1 b_4, \\ &\quad a_5 b_1 + a_2 b_3 + a_3 b_2 + a_1 b_5, a_6 b_1 + a_3 b_3 + a_1 b_6] \,. \end{aligned}$$

For different dimensions the stucture of the multiplication operator changes essentially. Therefore, the multiplication cannot be described independently of the dimension and the evaluation order, it must be determined by a specific rule.

[9]In the *combinat*-package the routines *inttovec* (shortly for 'integer to vector') resp. *vectoint* (shortly for 'vector to integer').

As an example for the differentiation consider the derivative of Q_a with respect to the variable x_1 :

$$\partial Q_a / \partial x_1 = a_2 + 2a_4 x_1 + a_5 x_2 \equiv [a_2, 2a_4, a_5, 0, 0, 0, 0] \ .$$

In general, the differentiation operation reduces the maximum order of a coefficient representation by one.

The substitution of variables by polynomials of new variables is the most complex operation. The explicit formulas are left out here for brevity.

For the inclusion into the whole concept a modular structure is useful. Two basic modules are implemented for the conversion of multivariate functions to the corresponding coefficient representations and vice versa. These two modules are the basis for the calculation of the multiplication and substitution operators.

Operator modules are only calculated if they do not exist already in a data base. So multiple calculations are avoided. This method can often be employed in computer algebra.

5.2 Treatment of Discontinuities

Many discontinuities $d(\boldsymbol{x})$ in engineering systems can be expressed in the following general form:

$$d(\boldsymbol{x}) = \begin{cases} h(\boldsymbol{x}) & \text{for} \quad s(\boldsymbol{x}) \geq 0 \\ -h(\boldsymbol{x}) & \text{for} \quad s(\boldsymbol{x}) < 0 \end{cases} \ . \tag{5.1}$$

The scalar function $h(\boldsymbol{x})$ describes the non-smooth quantity while the scalar function $s(\boldsymbol{x})$ stands for the switching criterion. Both functions can depend on the complete state vector $\boldsymbol{x}$.

From this formulation two problems arise:

1. The governing equations are not given in a closed symbolic form so that it is very difficult to apply analytical algorithms.
2. The need for partial derivatives in the nonlinear stability analysis of the dynamic equations produces problems.

Both drawbacks can be overcome by introducing a smooth approximation $k(\boldsymbol{x})$ of the function $d(\boldsymbol{x})$ in terms of a suitable inverse tangent function:

$$d(\boldsymbol{x}) \approx k(\boldsymbol{x}) = h(\boldsymbol{x}) \frac{2}{\pi} \lim_{c \to \infty} \arctan c\, s(\boldsymbol{x}) \ . \tag{5.2}$$

In Fig. 5.1 the normalized inverse tangent function

$$\kappa\,(x) = \frac{2}{\pi}\;\arctan c\,x \tag{5.3}$$

is plotted for different values of the numerical parameter c. Clearly it can be seen that this function provides the required sign change at the origin. The accuracy increases with increasing c.

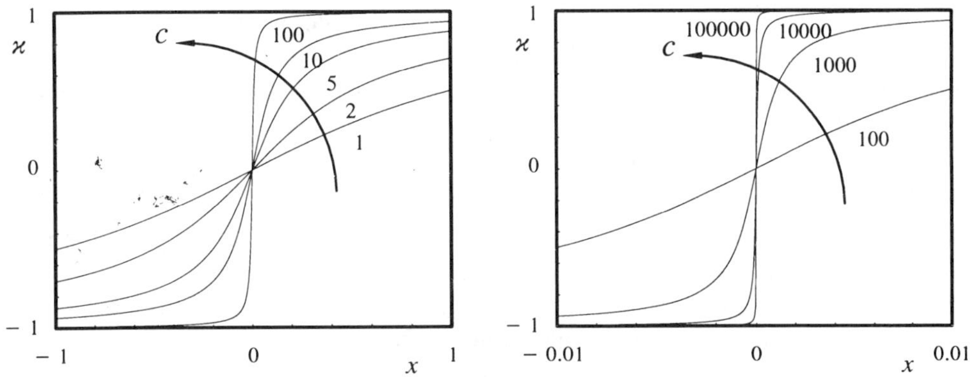

Figure 5.1: Normalized inverse tangent function for different values of c

The first derivative of $k\,(\boldsymbol{x})$ with respect to x_j becomes

$$\frac{\partial k(\boldsymbol{x})}{\partial x_i} = \frac{\partial h(\boldsymbol{x})}{\partial x_i}\,\frac{2}{\pi}\,\lim_{c\to\infty}\;\arctan c\,s(\boldsymbol{x}) + \mathcal{R}_1 \tag{5.4}$$

with

$$\mathcal{R}_1 = h(\boldsymbol{x})\,\frac{\partial s(\boldsymbol{x})}{\partial x_i}\,\lim_{c\to\infty}\,\frac{2\,c}{\pi\,(1+c^2\,s(\boldsymbol{x})^2)}\;. \tag{5.5}$$

The second derivative of $k\,(\boldsymbol{x})$ with respect to x_i and x_j is given by

$$\frac{\partial^2 k(\boldsymbol{x})}{\partial x_i\,\partial x_j} = \frac{\partial^2 h(\boldsymbol{x})}{\partial x_i\,\partial x_j}\,\frac{2}{\pi}\,\lim_{c\to\infty}\;\arctan c\,s(\boldsymbol{x}) + \mathcal{R}_2 \tag{5.6}$$

with

$$\mathcal{R}_2 = \left(\frac{\partial h(\boldsymbol{x})}{\partial x_i}\,\frac{\partial s(\boldsymbol{x})}{\partial x_j} + \frac{\partial h(\boldsymbol{x})}{\partial x_j}\,\frac{\partial s(\boldsymbol{x})}{\partial x_i} + h(\boldsymbol{x})\,\frac{\partial^2 s(\boldsymbol{x})}{\partial x_i\,\partial x_j}\right)\lim_{c\to\infty}\,\frac{2\,c}{\pi\,(1+c^2\,s(\boldsymbol{x})^2)}$$

$$-h(\boldsymbol{x})\,s(\boldsymbol{x})\,\frac{\partial s(\boldsymbol{x})}{\partial x_i}\,\frac{\partial s(\boldsymbol{x})}{\partial x_j}\,\lim_{c\to\infty}\,\frac{4\,c^3}{\pi\,\left(1+c^2\,s(\boldsymbol{x})^2\right)^2}\;. \tag{5.7}$$

Analogously, the third derivative of $k(\boldsymbol{x})$ with respect to x_i, x_j, and x_k has the same structure:

$$\frac{\partial^3 k(\boldsymbol{x})}{\partial x_i\,\partial x_j\,\partial x_k} = \frac{\partial^3 h(\boldsymbol{x})}{\partial x_i\,\partial x_j\,\partial x_k}\,\frac{2}{\pi}\lim_{c\to\infty}\arctan c\,s(\boldsymbol{x}) + \mathcal{R}_3\,. \tag{5.8}$$

The explicit expression for $\mathcal{R}_3$ is rather involved and is omitted for brevity.

It can be seen that the analytic calculation of the partial derivatives of the smooth approximation (5.2) becomes more and more complex the higher the derivative. Another problem is the appearance of powers of the numerical parameter c in the remainders $\mathcal{R}_i$ leading to numerical problems during the evaluation of the expressions because the parameter c has to be chosen as large as possible in order to guarantee a good approximation.

These observations lead to the concept of passive and active state variables: the main point is that the quantities $\mathcal{R}_i$ — if no jumps occur — are of the order $\mathcal{O}\left[c^{-1}\right]$ and therefore approach zero with $c \to \infty$. This means that for this case they can be neglected for the calculation of the partial derivatives.

For the efficient calculation of the partial derivatives of (5.2) it is thus advantageous to distinguish between active and passive state variables:

$$d(\boldsymbol{x}) \approx k(\boldsymbol{x},\tilde{\boldsymbol{x}}) = h(\boldsymbol{x})\,\frac{2}{\pi}\lim_{c\to\infty}\arctan c\,s(\tilde{\boldsymbol{x}})\,. \tag{5.9}$$

Passive variables $\tilde{\boldsymbol{x}}$

- always have the same value as the corresponding active variables $\boldsymbol{x}$,
- are treated as constants for the calculation of the derivatives,
- preserve smoothness of the partial derivatives.

This means that for the calculation of the partial derivatives the inverse tangent function is treated as a non-constant coefficient which preserves all smoothness properties but does not increase the complexity of the already rather involved expressions.

The partial derivatives with respect to the state variables are thus given in the simplified form

$$\frac{\partial^n k(\boldsymbol{x},\tilde{\boldsymbol{x}})}{\partial x_i\cdots} = \frac{\partial^n h(\boldsymbol{x})}{\partial x_i\cdots}\,\frac{2}{\pi}\lim_{c\to\infty}\arctan c\,s(\tilde{\boldsymbol{x}})\,. \tag{5.10}$$

This concept is well-suited for a computer algebra implementation.

6 Case Study: Submerged Double Pendulum

Remotely operated underwater vehicles are increasingly important tools for inspection and maintenance tasks e.g. in off-shore engineering and deap sea mining. A typical configuration of the system is shown in Fig. 6.1.

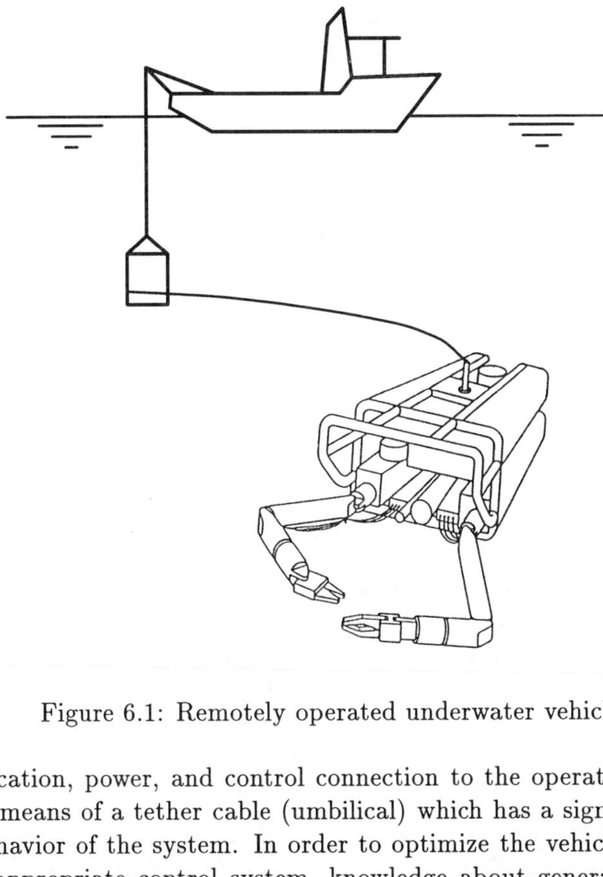

Figure 6.1: Remotely operated underwater vehicle

The communication, power, and control connection to the operation basis is usually performed by means of a tether cable (umbilical) which has a significant influence on the overall behavior of the system. In order to optimize the vehicle construction and to design an appropriate control system, knowledge about general characteristics of the dynamics of the system is required. Therefore, the cable needs to be included into the model as well. A promising approach is to model the cable as a system of interconnected rigid cylindrical elements, BEVILACQUA ET AL. (1991). A sufficiently accurate description of the hydrodynamic coupling (added mass effects and hydrodynamic damping) of the cable elements to the surrounding medium water is of major importance.

For the investigation of the basic properties of such a system the cable model is reduced to a simple submerged planar double pendulum and a mathematical model including all major effects is developed and analyzed. The results could serve as a basis for the description of the entire cable as well as for a mechanical sensor system for measuring the distance between the ROV and an offshore structure.

For the system under consideration subjected to periodic excitation the approach for a systematic computer-aided stability analysis is exemplarily carried out for a specific parameter set. A special difficulty is given by the non-smooth damping characteristics according to *Morison's equation* for the hydrodynamic coupling. Therefore, the obtained model equations are not given in a closed symbolic form. Furthermore, the non-smoothness leads to discontinuities in the second derivatives of the dynamic equations.

Therefore, the damping forces are included smoothly and in a closed symbolic form as is discussed in Section 5.2. This method leads to a drastically increased complexity of the model equations and computer algebra is absolutely necessary for the development as well as for the analysis.

6.1 Computer-Aided Modeling

A simple mechanical model of the planar submerged inverted double pendulum under consideration is given in Fig. 6.2.

The length of the bars is denoted by l_1 and l_2, the distance between a joint and the mass center C_i is given by s_1 and s_2, and the distance between the mass center C_i and the buoyancy center B_i by a_1 and a_2, respectively.

Applied forces acting at the two bars are buoyancy forces (A_1, A_2), gravity forces (G_1, G_2), and hydrodynamic forces (H_{x2}, H_{y1}, H_{y2}). Furthermore, the hydrodynamic force H_j is introduced acting at the joint between the two bars. From the mechanical point of view this joint belongs to the first bar and has diameter d_j and length l_j (in z–direction). Hydrodynamic torques with respect to the mass center are given by M_1 and M_2. The excitation is characterized by the angular amplitude a and frequency ω and is transmitted through a torsional spring with constant k_s.

The upright position of this two degrees-of-freedom rigid body system is assumed to be a stable equilibrium position due to the buoyancy forces acting at the bars. For the mathematical description the absolute angles α_1 and α_2 are chosen as generalized coordinates which are summarized in the position vector $\boldsymbol{y} = [\alpha_1 \;\; \alpha_2]^T$.

The dynamic coupling of the system to the surrounding water must be described as well. Specifically, the following effects are to be considered:

- added mass and inertia terms due to the simultaneous acceleration of the water,
- quadratic damping effects due to the viscosity of the surrounding medium.

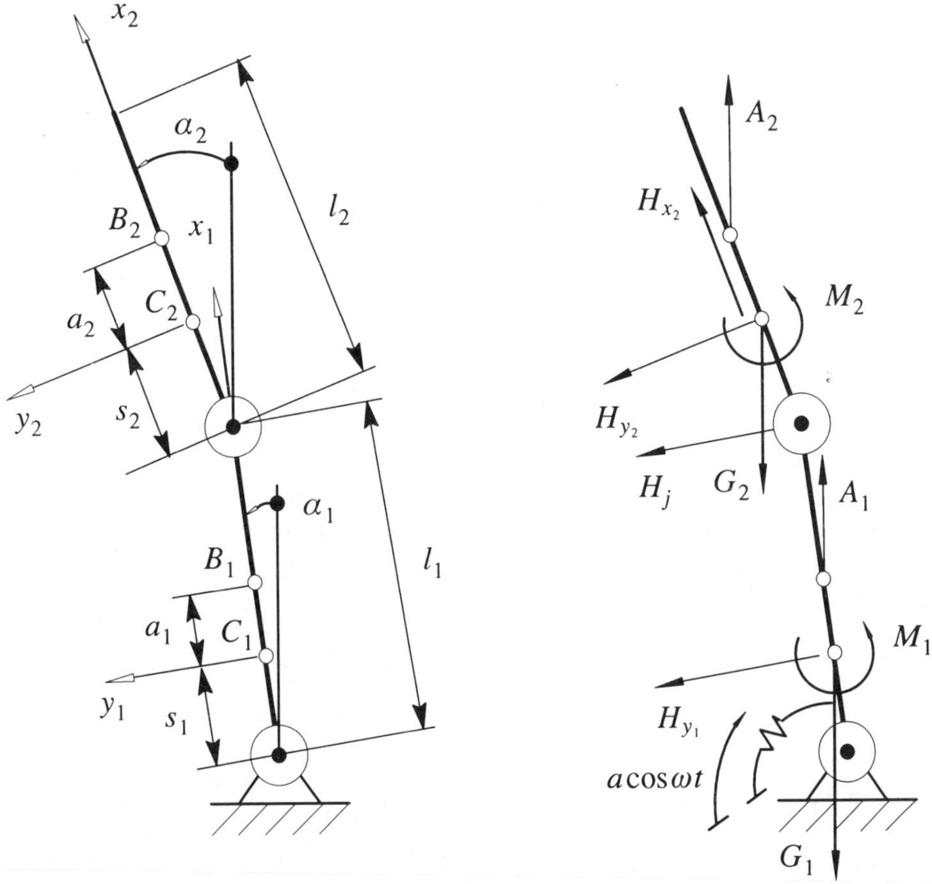

Figure 6.2: Mechanical model of the submerged double pendulum
a) Geometry b) Forces and excitation

An exact description of the underlying fluid mechanical phenomena by means of partial differential transport equations for the continuous medium leads to major theoretical and practical difficulties for the formulation as well as for the solution. Therefore, it is customary to utilize an approximate approach as is presented in the following.

Added Mass Effects

Inertia properties of moving bodies in a continuous medium like water are significantly increased because surrounding fluid particles are also accelerated. This effect can be taken into account by introducing a so-called *added mass matrix*, e.g. MEI (1983). Hence, the total inertia matrix $\bar{M}^e$ for a free element (mass m, inertia parameters

I_{xx} , I_{yy} , I_{zz} , I_{xy} , I_{xz} , I_{yz}) is given by the sum of the element's inertia matrix and the added mass matrix:

$$\bar{\boldsymbol{M}}^e = \begin{bmatrix} m & 0 & 0 & 0 & 0 & 0 \\ 0 & m & 0 & 0 & 0 & 0 \\ 0 & 0 & m & 0 & 0 & 0 \\ 0 & 0 & 0 & I_{xx} & I_{xy} & I_{xz} \\ 0 & 0 & 0 & I_{xy} & I_{yy} & I_{yz} \\ 0 & 0 & 0 & I_{xz} & I_{yz} & I_{zz} \end{bmatrix} + \begin{bmatrix} A_{11} & A_{12} & \cdots & A_{16} \\ A_{21} & A_{22} & & A_{26} \\ \vdots & & \ddots & \vdots \\ A_{61} & A_{62} & \cdots & A_{66} \end{bmatrix} .$$

Usually, only 21 significant added mass coefficients are considered because the added mass matrix is often assumed to be symmetric. Strict symmetry, however, applies only under certain additional conditions, FALTINSEN (1990).

For geometrically simply shaped bodies like the slender cylindrical bars of the double pendulum there are exact solutions available under the assumption of an idealized fluid NEWMAN (1977). For a body-fixed coordinate system with the origin located at the mass center C_i of the i–th body and the x_i–axis directed tangentially to the bar the only non-vanishing added mass coefficients in the case of a planar motion in the x_i-y_i–plane are

$$m_i^\star \equiv A_{22_i} = \rho_W\, V_i \;, \quad I_i^\star \equiv A_{66_i} = (\frac{1}{3}\, l_i^2 - s_i\, l_i + s_i^2)\, \rho_W\, V_i$$

where the tube volume is denoted by V_i and the water density by ρ_W .

Damping Forces

The hydrodynamic damping forces acting at the two bars of the pendulum can be calculated from the velocity distribution relative to the surrounding medium.

The relative velocity $\boldsymbol{u}$ must be split into a normal component $u_y(x)$ and a tangential component u_x to account for the different damping coefficients for the respective (local) directions. By assuming a body fixed frame as was introduced before, the relative velocity field can be found from kinematical considerations and is given for the first bar by

$$\begin{bmatrix} u_{x_1} \\ u_{y_1}(x_1) \end{bmatrix} = \begin{bmatrix} 0 \\ -\dot{\alpha}_1\,(x_1 + s_1) \end{bmatrix} \tag{6.1}$$

and for the second bar by

$$\begin{bmatrix} u_{x_2} \\ u_{y_2}(x_2) \end{bmatrix} = \begin{bmatrix} -l_1\,\dot{\alpha}_1\,\sin(\alpha_2 - \alpha_1) \\ -l_1\,\dot{\alpha}_1\,\cos(\alpha_2 - \alpha_1) - \dot{\alpha}_2\,(x_2 + s_2) \end{bmatrix} . \tag{6.2}$$

According to Morison's formula, the hydrodynamic damping force $\boldsymbol{H}$ and torque M for an infinitesimal slice of length dx take the form

$$\begin{bmatrix} dH_x \\ dH_y(x) \end{bmatrix} = \frac{1}{2}\rho_W \, d_t \begin{bmatrix} c_t \, u_x \, |u_x| \\ c_n \, u_y \, |u_y| \end{bmatrix} dx \;, \tag{6.3}$$

$$dM(x) = dH_y \, x \tag{6.4}$$

where the tube diameter is denoted by d_t. The dimensionless damping coefficients c_t and c_n must be determined from experiments. Approximate values can be found e.g. in HWANG (1986).

An integration over the entire element length yields the resulting damping force and torque acting at the considered pendulum tube. A smooth approximation $k_u(u)$ of the damping function $u\,|u|$ in terms of the inverse tangent function

$$u\,|u| \approx k_u(u) = u^2 \lim_{c\to\infty} \frac{2}{\pi} \arctan cu \tag{6.5}$$

is used for Morison's approach. Then, after substitution of the variable u with the explicit kinematical expressions given before, all integrations can be performed analytically and the results are obtained in a closed symbolic and differentiable form. The whole derivation is performed via symbolic computation.

Generation of the Dynamic Equations and State Space Representation

For multibody system models the generation of the symbolic equations of motion can be performed by the symbol manipulation program Neweul, KREUZER AND LEISTER (1991). The computer-aided methodology to obtain the model equations of a fluid-coupled system is shown in Fig. 6.3. Starting point is the geometrical configuration of the system according to Fig. 6.2. This can be described by means of position vectors $\boldsymbol{r}_i$ and rotation matrices $\boldsymbol{S}_i$ of the two bodies. The application of the special-purpose computer algebra package Neweul results in the dynamic equations of the system where added mass terms are not included because Neweul is not designed for this special case.

It is possible to account for the added mass terms by making use of the kinematics output of Neweul which is represented in terms of the Jacobian matrices for translation and rotation for each body. The general-purpose computer algebra system Maple is used to construct the global Jacobian matrix $\bar{\boldsymbol{J}}$ and the global inertia matrix $\bar{\boldsymbol{M}} = \mathrm{diag}\{\bar{\boldsymbol{M}}_i^e\}$ of the system and to perform the necessary transformation

$$\boldsymbol{M} = \bar{\boldsymbol{J}}^T \bar{\boldsymbol{M}} \bar{\boldsymbol{J}} \tag{6.6}$$

to obtain the complete generalized inertia matrix $\boldsymbol{M}$ including the added mass terms.

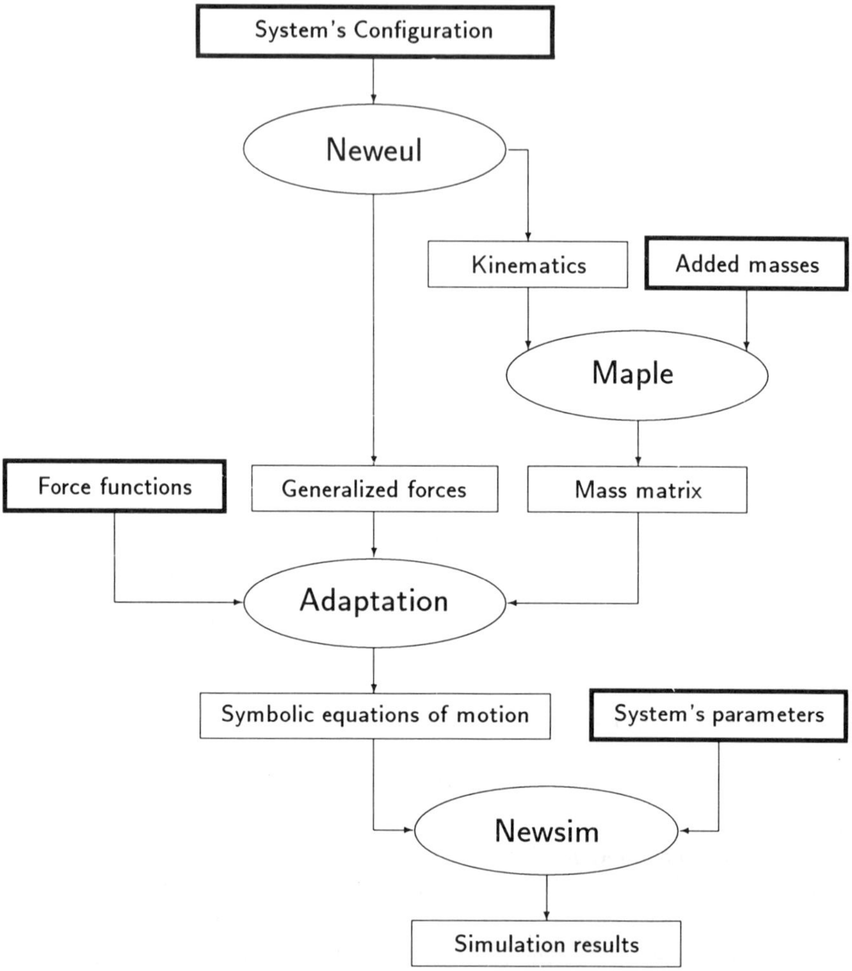

Figure 6.3: Generation of the dynamic equations

All applied forces and torques acting on the system must be user-provided in terms of functions of the state variables and the time. This results in the symbolic vector differential equation of motion

$$\boldsymbol{M}(\boldsymbol{y},t)\,\ddot{\boldsymbol{y}}+\boldsymbol{k}\,(\boldsymbol{y},\dot{\boldsymbol{y}},t)=\boldsymbol{q}\,(\boldsymbol{y},\dot{\boldsymbol{y}},t)$$

as it was already introduced in Eq. (2.24). Specifically, for the submerged double pendulum the matrices take the following form:

$$\boldsymbol{M} = \begin{bmatrix} s_1^2\,(m_1+m_1^\star)+l_1^2(m_2+\frac{1}{2}m_2^\star)+ & l_1\,s_2\,\cos(\alpha_2-\alpha_1)\,(m_2+m_2^\star) \\ \frac{1}{2}l_1^2 m_2^\star\cos 2(\alpha_2-\alpha_1)+I_1+I_1^\star & \\ l_1\,s_2\,\cos(\alpha_2-\alpha_1)\,(m_2+m_2^\star) & s_2^2\,(m_2+m_2^\star)+I_2+I_2^\star \end{bmatrix},$$

$$\boldsymbol{k} = \begin{bmatrix} -m_2\,l_1\,s_2\,\dot\alpha_2^2\sin(\alpha_2-\alpha_1) \\ m_2\,l_1\,s_2\,\dot\alpha_1^2\sin(\alpha_2-\alpha_1) \end{bmatrix},$$

$$\boldsymbol{q} = \begin{bmatrix} [G_1\,s_1 - A_1\,(s_1+a_1)]\sin\alpha_1 + (G_2-A_2)l_1\sin\alpha_1 + H_{y1}(\boldsymbol{y},\dot{\boldsymbol{y}})\,s_1 + \\ H_{x2}(\boldsymbol{y},\dot{\boldsymbol{y}})l_1\sin(\alpha_2-\alpha_1) + H_{y2}(\boldsymbol{y},\dot{\boldsymbol{y}})\,l_1\cos(\alpha_2-\alpha_1) + \\ H_j(\boldsymbol{y},\dot{\boldsymbol{y}})\,l_1 + M_1 + k_s(a\cos\omega t - \alpha_1) \\ [G_2\,s_2 - A_2\,(s_2+a_2)]\sin\alpha_2 + H_{y2}(\boldsymbol{y},\dot{\boldsymbol{y}})\,s_2 + M_2 \end{bmatrix}.$$

Here, hydrodynamic force and torque terms are not explicitly expressed as functions of $\boldsymbol{y}$ and $\dot{\boldsymbol{y}}$ because this would increase the length of the expressions considerably. The simulation package Newsim is designed to directly compute the numerical solution of these model equations for an arbitrary parameter set.

Via Maple the obtained second order system can easily be transformed into the standard state space representation (1.1) of dynamic systems with the state vector $\boldsymbol{x} = [\,\boldsymbol{y}^T \;\; \dot{\boldsymbol{y}}^T\,]^T$. This formulation is the starting point for the stability analysis described in the next section.

6.2 Computer-Aided Stability Analysis

The nonlinear stability analysis for the double pendulum is performed according to the steps shown in Fig. 1.1. The parameter values used for the investigation are given in Table 6.1.

Table 6.1: Parameter values

m_1	7.31	[kg]	m_2	0.88	[kg]
$m_1^\star$	2.12	[kg]	$m_2^\star$	0.43	[kg]
I_1	0.17	[kg m^2]	I_2	0.018	[kg m^2]
$I_1^\star$	0.22	[kg m^2]	$I_2^\star$	0.010	[kg m^2]
A_1	28.5	[N]	A_2	4.2	[N]
l_1	0.4	[m]	l_2	0.4	[m]
s_1	0.061	[m]	s_2	0.069	[m]
a_1	0.21	[m]	a_2	0.081	[m]
d_j	0.09	[m]	l_j	0.29	[m]
d_t	0.032	[m]	k_s	236.5	[Nm/rad]
c_n	1.26	[-]	c_t	0.1	[-]
ρ_W	1000.	[kg/m^3]	ω	$2\,\pi$	[rad/s]

By making use of the concept of passive and active variables and by utilizing structural properties of mechanical systems as was discussed before, the calculation of the partial derivatives of the model equations could be reduced by a remarkable amount.

Applying the analysis procedure for a stable one-periodic solution of the double pendulum a symmetry-breaking pitchfork bifurcation with critical eigenvalue $\lambda^\star = 1$ and a period-doubling flip bifurcation with critical eigenvalue $\lambda^\star = -1$ could be detected. On the basis of the symbolically given map a very accurate estimation of the bifurcation parameter values $a^\star$ of the excitation amplitude a could be obtained. Exemplarily, the iteration sequence for the pitchfork bifurcation is given in Table 6.2.

Table 6.2: Iteration sequence for pitchfork bifurcation parameter value

Iteration	a	λ
0	0.1000000	0.48912898 + 0.32400075 j
1	0.1407018	0.84036745
2	0.1605386	0.99358588
3	0.1617132	1.00161352
4	0.1614095	0.99954580
5	0.1614944	1.00012418
10	0.1614761	0.99999996

It is interesting to note that the algorithm converges even though the periodic solution is unstable for $a > a^\star$.

After this symmetry-breaking bifurcation took place the symmetric type of motion of the double pendulum changes to an unsymmetric one. This is illustrated in Fig. 6.4 where the extreme positions (turning points) of the pendulum bars are shown for the two qualitatively different situations.

Examining the normal form of the reduced (critical) system clearly reveals a stable supercritical pitchfork bifurcation and an unstable subcritical flip bifurcation.

Numerical simulations of the steady-state behavior of the double pendulum for control parameter values chosen close to the critical bifurcation values confirm the obtained results as can be seen in the bifurcation and phase diagrams for the second bar given in Fig. 6.5.

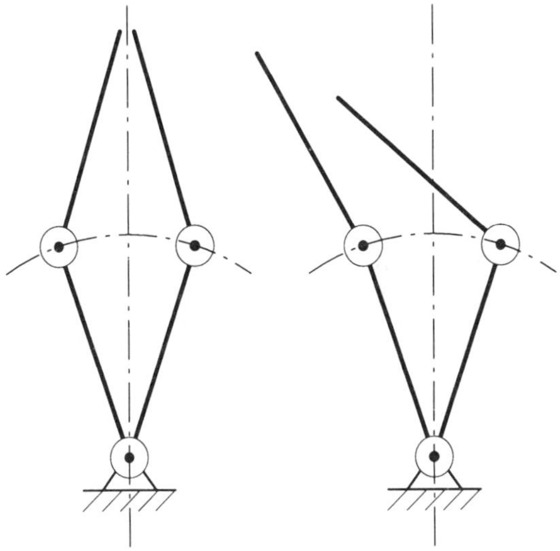

Figure 6.4: a) Symmetric motion b) Unsymmetric motion

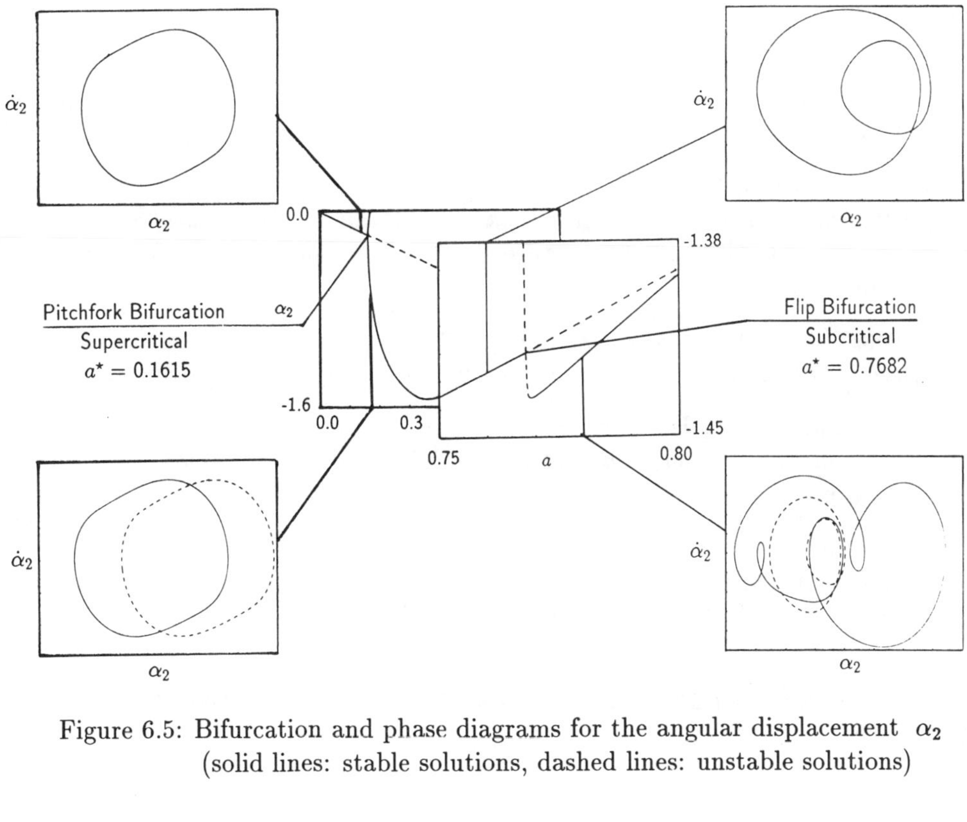

Figure 6.5: Bifurcation and phase diagrams for the angular displacement α_2 (solid lines: stable solutions, dashed lines: unstable solutions)

Literature

ARNOL'D, V.I., 1987,
Geometrische Methoden in der Theorie der gewöhnlichen Differentialgleichungen, Basel: Birkhäuser.

BEVILACQUA, L., KLECZKA, W., AND KREUZER, E., 1991,
On the Mathematical Modeling of ROV'S, Proc. of the Symposium on Robot Control (Vienna, Austria, 1991), Troch, I., Desoyer, K., and Kopacek, P. (eds.), Vienna: ÖWPZ, pp. 595 - 598.

CARR, J., 1981,
Applications of Centre Manifold Theory, New York/...: Springer.

CHAR, B.W., GEDDES, K.O., GONNET, G.H., LEONG, B.L., MONAGAN, M.B., AND WATT, S.M., 1991,
Maple V *Language Reference Manual*, Berlin/...: Springer.

ELPHIK, C., BRACHET, M.E., AND TIRAPEGUI, E., 1987,
A simple global characterization for normal forms of singular vector fields, Physica D 29, pp. 95-127.

FALTINSEN, O.M., 1990,
Sea loads on ships and offshore structures, Cambridge/...: Cambridge University Press.

FREIRE, E., GAMERO, E., PONCE, E., AND G.-FRANQUELO, L., 1988,
An Algorithm for Symbolic Computation of Center Manifolds, In: *Symbolic and Algebraic Computation*, P. Gianni (ed.), Lecture Notes in Computer Science Vol. 358, New York/...: Springer.

GOLUBITSKY, M., AND SCHAEFFER, D.G., 1985,
Singularities and Groups in Bifurcation Theory, Vol. 1, New York/...: Springer.

GUCKENHEIMER, J., AND HOLMES, P., 1986,
Nonlinear Oscillations, Dynamical Systems, and Bifurcations of Vector Fields, New York/...: Springer.

HAACK, C., 1990,
Rechnergestützte Verfahren zur Analyse nichtlinearer dynamischer Probleme, Stuttgart: Universität Stuttgart, Institut B für Mechanik, Diplomarbeit DIPL-28.

HSU, C.S., 1980,
A Theory of Cell-to-Cell Mapping Dynamical Systems, J. of Applied Mechanics 47, pp. 931 - 939.

HSU C.S., KREUZER, E., AND KIM, M.C., 1990,
Bifurcation characteristics of piecewise linear mappings and their implications, In: Dynamics and Stability of Systems, Vol. 5, No. 4, Oxford University Press.

HWANG, Y.-L., 1986,
Nonlinear dynamic analysis of mooring lines, Proc. of OMAE Symp., Vol. 3, No. 5, pp. 499 – 506.

KLECZKA, M., 1985,
Zur Berechnung der Ljapunov-Exponenten und deren Bedeutung, Stuttgart: Universität Stuttgart, Institut B für Mechanik, Studienarbeit STUD-16.

KLECZKA, M., 1991,
Methoden zur Verzweigungsanalyse mit Anwendung auf einen Spielschwinger, Fortschrittberichte VDI, Reihe 11, Nr. 153, Düsseldorf: VDI-Verlag.

KLECZKA, M., KLECZKA, W., AND KREUZER, E., 1990,
Bifurcation Analysis: A Combined Numerical and Analytical Approach, Proc. of the Workshop on Cont. and Bif. (Leuven, Belgium, 1989), Roose, D., DeDier, B., and Spence, A. (eds.), Dordrecht: Kluwer Acad. Publ. (NATO ASI series), pp. 123 – 137.

KLECZKA, W., 1989,
Einsatz von Computer-Algebra zur Analyse nichtlinearer dynamischer Probleme, Stuttgart: Universität Stuttgart, Institut B für Mechanik, Studienarbeit STUD-50.

KLECZKA, W., AND KREUZER, E., 1991,
Systematic Computer-Aided Analysis of Dynamic Systems, Proc. of the International Symposium on Symbolic and Algebraic Computation (Bonn, F.R.G., 1991), Stephen M. Watt (ed.), New York: ACM Press, pp. 429 – 430.

KREUZER, E., 1987,
Numerische Untersuchung nichtlinearer dynamischer Systeme, Berlin/...: Springer.

KREUZER, E., AND LEISTER, G., 1991,
Programmpaket Neweul*'90*, Stuttgart: Universität Stuttgart, Inst. B für Mechanik, Anleitung AN-23.

KREUZER, E., AND SCHIEHLEN, W., 1990,
Neweul-*Software for the Generation of Symbolical Equations of Motion*, In: *Multibody Systems Handbook*, Schiehlen, W. (ed.), Berlin/...: Springer, pp. 181 – 202.

LICHTENBERG, G., 1992,
Symbolische Berechnung von Normalformen für nichtlineare dynamische Systeme, Hamburg: Technische Universität Hamburg-Harburg, Arbeitsbereich Meerestechnik II, Diplomarbeit DIPL-4.

MEI, C.C., 1983,
The Applied Dynamics of Ocean Surface Waves, New York/...: John Wiley & Sons.

NEWMAN, J.N., 1977,
Marine Hydrodynamics, Cambridge, Mass.: MIT Press.

PONCE, E., GAMERO, E., AND FREIRE, E., 1991,
Normal forms for planar systems with nilpotent linear part, In: *Bifurcation and Chaos: Analysis, Algorithms, Applications*, Basel: Birkhäuser.

RAND, R.H., AND ARMBRUSTER, D., 1987,
Perturbation Methods, Bifurcation Theory and Computer Algebra, New York/...: Springer.

SEYDEL, R., 1988,
From Equilibrium to Chaos: Practical Bifurcation and Stability Analysis, New York: Elsevier.

TROGER, H., AND STEINDL, A., 1991,
Nonlinear stability and bifurcation theory, Wien/...: Springer.

WIGGINS, S., 1990,
Introduction to Applied Nonlinear Dynamical Systems and Chaos, New York/...: Springer.

WILMERS, W., 1988,
Verzweigungsphänomene in mechanischen Oszillatoren, Stuttgart: Universität Stuttgart, Institut B für Mechanik, Diplomarbeit DIPL-24.

ZURMÜHL, R., AND FALK, S., 1984,
Matrizen und ihre Anwendungen; Teil 1: Grundlagen, Berlin/...: Springer.